PETRI NET SYNTHESIS FOR DISCRETE EVENT CONTROL OF MANUFACTURING SYSTEMS

THE KLUWER INTERNATIONAL SERIES IN ENGINEERING AND COMPUTER SCIENCE

DISCRETE EVENT DYNAMIC SYSTEMS

Consulting Editor

Yu-Chi Ho
Harvard University

GRADIENT ESTIMATION VIA PERTURBATION ANALYSIS, P. Glasserman
ISBN: 0-7923-9095-4

PERTURBATION ANALYSIS OF DISCRETE EVENT DYNAMIC SYSTEMS, Yu-Chi Ho and Xi-Ren Cao
ISBN: 0-7923-9174-8

PETRI NET SYNTHESIS FOR DISCRETE EVENT CONTROL OF MANUFACTURING SYSTEMS

By

MengChu Zhou
New Jersey Institute of Technology

Frank DiCesare
Rensselaer Polytechnic Institute

KLUWER ACADEMIC PUBLISHERS
BOSTON/DORDRECHT/LONDON

Distributors for North America:
Kluwer Academic Publishers
101 Philip Drive
Assinippi Park
Norwell, Massachusetts 02061 USA

Distributors for all other countries:
Kluwer Academic Publishers Group
Distribution Centre
Post Office Box 322
3300 AH Dordrecht, THE NETHERLANDS

Library of Congress Cataloging-in-Publication Data

Zhou, MengChu
Petri net synthesis for discrete event control of manufacturing systems / MengChu Zhou. Frank DiCesare.
p. cm. -- (The Kluwer international series in engineering and computer science. Discrete event dynamic systems)
Includes bibliographical references and index.
ISBN 0-7923-9289-2 (alk. paper)
1. Flexible manufacturing systems--Mathematical models. 2. Petri nets. 3. Production control--Mathematical models. I. DiCesare, Frank. II Title. III Series.
TX155.6.Z48 1993
670.42'7'015118--dc20 92-36721
CIP

Printed on acid-free paper.

Printed in the United States of America

To our parents, wives, children, and to the memories of Si Ping Zhou and Francesco DiCesare.

Contents

List of Figures

List of Tables

Notation

The following notation and abbreviations are used throughout the book.

Section 1.2.3 means Chapter 1, Section 2, Subsection 3, likewise for example, theorem, lemma, figure, and table.

0_n = an n-dimensional vector with each component being zero
1_n = an n-dimensional vector with each component being one
$0_{n \times n}$ = an $n \times n$-dimensional matrix with each component being zero
A-place = a place that represents operation for most cases
AGV = Automated guided vehicle
AMTS = Automated material transfer system
ASRS = Automated storage and retrieval system
B-place = a place that represents status of resources whose quantity is relatively fixed
b = the capacity of a buffer
b_i = the capacity of Buffer i for $i = 1, 2, \ldots$
C-place = a place that represents status of resources whose quantity varies
$C(t, t')$ = the token capacity between transitions t and t' in a Petri net
C^i = the token capacity between the ith PME and $(i + 1)$st PME in an SME
CAD = Computer aided design
D = a set of pairs of transitions in a PME, and so is D^i for $i = 1, 2, \ldots$
$D^{(n)}$ = an n-PME or an SME containing n PMEs
DEDS = Discrete event dynamic system
DEM = Discrete event Model

DES = Discrete event system

$\varepsilon : P \rightarrow N = \{0, 1, ...\}$ and is called the error information function (also called the ε-function)

$EC(x)$ = an elementary circuit starting and ending with x and is used as a set of nodes

EDF = Essentially decision free

$EP(x, y)$ = one elementary path, or the set of all elementary paths from x to y, and is used as a set of nodes, where $x, y \in P \cup T$.

$EP(x, Y)$ = the set of all elementary paths from x to any node in Y where Y is a set of nodes

FIFO = First-In-First-Out

FMS = Flexible manufacturing system

FRP = Finitely recursive process

$I : P \times T \rightarrow \{0, 1\}$ = an input function in a Petri net

I_n = the $n \times n$-dimensional unity matrix

$I_S : P_S \times T_S \rightarrow \{0, 1\}$ = an input function in a sub-Petri net

$J_n = \begin{pmatrix} 0_{n-1}^T & 1 \\ I_{n-1} & 0_{n-1} \end{pmatrix}$

$k_i = |D^i|$, the number of the pairs of transitions in a PME $D^i, i = 1, 2, ...$

L = the number of PMEs in an SME

MMS = Manufacturing message specification

M_i = Machine i for $i = 1, 2, ...$

$m : P \rightarrow N$ = the marking vector in a Petri net Z

$m = \begin{pmatrix} \mu \\ m(P_E) \end{pmatrix}$ in a net with a PME or an SME and is written as $(\mu, m(p_E))$ to save space

$m(p)$ = the number of tokens in place p

$m_0 : P \rightarrow N$ = the initial marking vector in a Petri net Z

$m_0' : P' \rightarrow N$ = the initial marking vector in a Petri net Z'

$N = \{0, 1, 2, ...\}$, the set of all natural numbers

$N^+ = \{1, 2, ...\}$, the set of all positive integers

$N_k = \{1, 2, ..., k\}$

$O : P \times T \rightarrow \{0, 1\}$ = an output function in a Petri net

$O_S : P_S \times T_S \rightarrow \{0, 1\}$ = an output function in a sub-Petri net

P = a set of places in a Petri net Z; and $|P|$ is the cardinality of the set P

P_A = a set of A-places

P_B = a set of B-places

P_C = a set of C-places

P_S = the set of places in a sub-Petri net

p = a place in a Petri net, and so are p', p_E, and p_i for $i = 1, 2, \ldots$
PLC = Programmable logic controller
PME = Parallel mutual exclusion
PN = Petri net
PNDL = Petri net description language
PNSS = Petri net supervisory system
Q = sub-Petri net R = the set of all non-negative real numbers
$R(Z, m_0)$ = the reachability set of a Petri net $Z = (P, T, I, O, m_0)$
$R(Z', m_0')$ = the reachability set of a Petri net $Z' = (P', T', I', O', m_0')$
S = sub-Petri net
S' = sub-Petri net
S^* = the associate Petri net of a sub-Petri net
SME = Sequential mutual exclusion
SPN = Stochastic Petri net
SPNP = Stochastic Petri Net Package
$\tau : P \rightarrow R = \{x, x$ is a non-negative real number
$\tau(p)$ (called the τ-function) = the elapsed time in place p in P
$\tau_{max} : P \rightarrow R$, and $\tau_{max}(p)$ is the maximum elapsed time in place p
T = a set of transitions in a Petri net
T_a, T_b = subsets of T and so are T_a^i and T_b^i for $i = 1, 2, \ldots$
T_S = the set of transitions in a sub-Petri net
t = a transition in a Petri net, and so are $t', t'', t_i, t_{ai}, t_{bi}, t_{aj}^i$, and t_{bj}^i for $i, j = 1, 2, \ldots$
W_i = the ith workstation, $i = 1, 2, \ldots$
$Z = (P, T, I, O, m_0)$ and is a marked Petri net, so are $Z', Z^e(t, t')$, and $Z^\alpha(p)$
Z^n is a Petri net with an n-PME or SME
μ = a partial marking vector of $m \in R(Z^n, m_0')$ except $m(p_E)$
$\alpha^i, \alpha, \beta^i, \beta, \gamma^i$ and γ describe the relationships among initial markings of C-places and transitions in an SME or GSME
σ = the synchronic distance between two transitions in a Petri net

Besides, if the marking m' is reachable from m by firing a transition t, we write $m[t> m'$. If m' is reachable from m by firing a sequence of transitions $t_a, t_b, \ldots, t_c$, we write $m[t_a t_b \ldots t_c> m'$.

Each of e, f, g, h, and r represents a sequence of fireable transitions, and so are e_i, f_i, g_i, h_i, and r_i for $i = 1, 2, \ldots$, unless stated otherwise;

$m[f>$ represents a marking resulting from the firing of f at m;
$m(p)[f>$ is the number of tokens in p at marking $m[f>$. $m[\lambda> m$ is assumed; where λ is an empty string;
$\sharp(f,t)$ is the number of times t appears in f;
$\sharp(f,T)$ is the number of times transitions in T appear in f;
The letters i, j, k, u, v are used as indices;
(t,p) means an arc from transition t to place p, and (p,t) one from p to t in particular contexts.

Preface

Petri nets were introduced by Carl A. Petri in 1962. Since then there has been a growing interest in the theory and applications of nets for the modeling and analysis of asynchronous concurrent systems. Included among the areas of application are computer systems with concurrent programming and multiprocessor systems, protocol design and verification in computer networks, and the modeling and control of flexible manufacturing systems. These all can be characterized as discrete event dynamic systems (DEDS), the subject of intense current research and the series in which this book resides. The complexity resulting from the inherent nonlinearity and state space dimension found in most of these systems leads to unusual difficulty in design and analysis. With improper design, these systems are prone to deadlock, overflow and degraded performance. This is complicated by the need for powerful modeling and analysis tools to detect and correct these problems.

Petri net researchers have made remarkable strides in the modeling, analysis, and control of DEDS. The original approach was to model and then to analyze. This ultimately limited the size or class of systems which could be considered. To address this problem both top-down (stepwise refinement) and bottom-up (modular composition) synthesis methods were developed. The purpose of this book is to extend this synthesis theory by directly addressing the problem of shared resource systems and providing a hybrid methodology which systematically includes both top-down and bottom-up design. A modeler correctly using this approach is guaranteed a system which is logically correct in terms of liveness (no deadlock), boundedness (no overflow) and reversibility (re-initializable). This theory is presented in the context of manufacturing systems with realistic examples to illustrate the applicability to a wide class of problems involving synthesis for control design as well as error recovery. Limitations of these methods are presented as well as ideas for further research. Performance analysis of discrete event systems using Petri nets is an important topic not covered in this book.

Organization

Chapter 1 introduces discrete event systems and Petri nets as one of several approaches to their modeling and design. Chapter 2 defines Petri nets, their properties, and presents a manufacturing example and a commonly-used modeling methodology. It then relates manufacturing system characteristics to Petri net properties and provides concepts necessary for the subsequent developments.

The concepts and theory dealing with shared resources are presented in Chapters 3 and 4. Chapter 3 introduces the parallel mutual exclusion which models a resource shared by distinct independent processes and Chapter 4 proposes the sequential mutual exclusion to model a resource shared by sequentially related processes. A sequential mutual exclusion is a sequential composition of parallel mutual exclusions. Both chapters present the theoretical results for the conditions under which a net containing these structures is still bounded, live, and reversible. The concepts and results are also extended to more generalized cases. These results are from [Zhou 91a].

In Chapter 5, a hybrid methodology for synthesis of Petri nets is presented. Basic design modules and their relations to manufacturing are introduced. The design procedure which incorporates the preceding theory follows a summary of theoretical results. Design of buffer models and first-level Petri net models for different systems are also included. Most of these results were presented in [Zhou 90b, e, f; Zhou 92a]. Chapter 6 illustrates the hybrid methodology with the synthesis of a Petri net model for an automated manufacturing system. This results in a specification of net structure and initial markings, which guarantees liveness, boundedness and reversibility.

In Chapter 7, a scaled physical simulator of a flexible manufacturing system is described and the design and implementation of its supervisory controller is discussed. This realistic example is presented to illustrate the level of real world complexity the methods are able to address. Portions of this work were first presented in [Zhou 90c, 92b].

Chapter 8 extends Petri net controller concepts to automatic error recovery. Four basic error recovery methods are studied in the context of extended Petri nets. Application of this theory guarantees the liveness, boundedness, and reversibility of a Petri net controller when augmented to accommodate various errors. This is illustrated using an example of a robotic piston insertion station. This work originally appeared in [Zhou 89a].

In Chapter 9, past development in the applications of Petri net theory to manufacturing automation is summarized and future research is discussed.

Acknowledgments

We acknowledge the long range financial support by the industrially sponsored Computer Integrated Manufacturing research program of the Center for Manufacturing Productivity and Technology Transfer at Rensselaer Polytechnic Institute. This book is an outgrowth of the the first author's doctoral dissertation at Rensselaer Polytechnic Institute. Acknowledgment is owed Rensselaer's Department of Electrical, Computer, and Systems Engineering from which the first author received his Ph.D. and in which the second author is a Professor. The first author also acknowledges support by the Separately Budgeted Research (SBR) Program at New Jersey Institute of Technology and the Center for Manufacturing Systems at NJIT where he has been a faculty member of the Electrical and Computer Engineering Department and Manufacturing Engineering Program since 1990. Our particular gratitude is extended to Dr. Leo Hanifin, former Director of the CMPTT at RPI, who believed in and promoted research in the area of Petri nets and manufacturing control.

The Petri Net group meetings with Drs. I. Koh, M. Jeng, A. Giua, H. Jungnitz, R. Al-Jaar, Professors A. A. Desrochers and D. Guo, and Mr. D. Y. Lee and D. Rudolph helped to stimulate the early ideas of this book. Thanks also go to Drs. X. Dong, F. Wang, and S. J. Chang for sharing their experience.

Special thanks should go to Professor Y. Zhuang who helped to prepare the numerous diagrams and some text in this book, and also Fang Chen, the first author's wife who helped convert portions of the text into Latex.

Finally, without the support from our family, especially Fang Chen and Alice, the authors' wives, we could not have finished this book. We thank them very much for their patience and love.

MengChu Zhou
New Jersey Institute of Technology
Newark, New Jersey

Frank DiCesare
Rensselaer Polytechnic Institute
Troy, New Work

Chapter 1

Introduction

1.1 Motivation

Event-driven systems are of increasing importance in today's world because they are growing in number, size, and sophistication. It is therefore imperative to have systematic design methodologies in order to achieve desireable performance and to avoid catastrophic failures. These systems include automated manufacturing systems, communication networks, computer operating systems, office information systems, and the like. They may be asynchronous and sequential, exhibiting many characteristics: concurrency, conflict, mutual exclusion, and non-determinism. These characteristics are very difficult to describe using traditional control theory, which deals with systems of continuous or synchronous discrete variables modeled by differential or difference equations. In addition, inappropriate control of the occurrence of events may lead to a system deadlock, capacity overflows, or may otherwise degrade system performance. These types of systems are called *discrete event systems (DES)* in this work. They are also called *discrete event dynamic systems (DEDS)* [Ho 87], *discrete event model* [Inan 88], or sometimes *discrete event processes* [Wonham 84, 87].

To help ground the discussion of event-driven systems, a simple manufacturing example is presented as follows. The system consists of two different machines, a robot, and a buffer, and is sketched in Fig. 1.1. Every part from the input storage must be processed by Machine 1 first and then by Machine 2 to produce a final product. The robot is used for unloading both machines, and the buffer is used to store intermediate parts. Machines 1 and 2 load themselves from the input part storage and the buffer, respectively. Once a machine starts work on a part, it cannot be interrupted until the work is complete. The following characteristics are embedded in this system: it is event-driven, asynchronous, and sequential; it exhibits concurrency, conflict, mutual exclusion, and non-determinism. More importantly, such a simple system can contain a system deadlock which may result from improperly triggering a particular sequence of events.

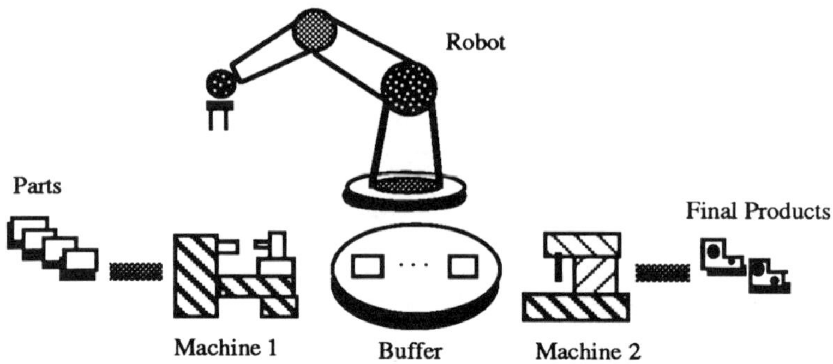

Figure 1.1: A discrete event system example

Event-driven: A discrete event system is characterized by a discrete state space where changes in state are triggered by event occurences. Precedence is a key relation between events, that is, any event may be dependent on the occurrence of other events. In the example shown in Fig.1.1, one event is the start of Machine 1 loading, fixturing and processing a part. The next event might be the end of this processing and the start of unloading of a part from Machine 1. Note that for a particular part the second event is dependent on the occurrence of the first. It may also be dependent on other events such as the release of the robot to make it available and the release of a buffer space.

Asynchronous: The asynchronous characteristic of discrete event systems is one of the most important properties by which they differ from traditional systems described by differential or difference equations. In time discretization of sampled systems, each change or step is synchronized by a global clock. In continuous systems, parameters vary continuously with time. However, in discrete event systems the events often occur asynchronously. For example, in the preceding discussion, the second event might be dependent on the ocurrence of three preceding events. The release of a buffer space will probably occur at a different time from the release of the robot or the end of processing on Machine 1. The start of unloading cannot occur until all of the preceding aynchronous events have occurred. Traditional methods fail to describe asynchronous systems.

Sequential Relation: Given a set of events, there may exist some sequential relationships among them. We say that there is a sequential relation between two events if one event can occur only after the occurrence of the other. Such a sequential relation can be described as a partial ordering. For the given example, *start_a_job* will occur first, followed by *complete_a_job* for the same part. Mathematically, there is a partial order between *start_a_job* and *complete_a_job*. A part can also be turned into a final product by Machines 1 and 2 in the system, but it has to be processed by M_1 first and then by M_2. This is another example of a sequential relation. In fact, given the same part, one set of events, *start_a_part_on_*M_2 and *complete_a_part_on_*M_2, has to follow another set of events, *start_a_part_on_*M_1 and *complete_a_part_on_*M_1.

Concurrency: By concurrency we mean that there are no sequential relationships among the concerned events. For example, physically, two events *complete_a_part_on_*M_1 and *complete_a_part_on_*M_2 are concurrent in the sense that either event may occur before the other. After the two machines start to work in the above system, they can operate concurrently since both can process their own parts at the same time after both are loaded by the robot. If machining processes are programmed into two independent programs which have been implemented on different computer processors, then they are executed concurrently or in parallel after they have been initiated at possibly different time moments. Also, either machine may break down at any time. One cannot tell exactly when such an unexpected event will occur. The events, M_1*_breaks_down* and M_2*_breaks_down*, have an uncertain sequential relationship. Thus we say they are concurrent.

Conflict: In the above system, there is a conflict when M_1 and M_2 complete their own parts and are ready to be unloaded, a buffer slot is ready for an intermediate part, and the robot is available. The robot can either unload M_1 or M_2, but not both of them simultaneously. When a resource is shared by different processes, such conflicts are often inevitable.

Mutual Exclusion: When the above conflict occurs, one can easily find that the two events *unload_*M_1 and *unload_*M_2 are mutually exclusive in the sense that they cannot occur at the same time, whereas after one is complete, the other can occur. Often in the literature, mutual exclusion among independent distributed processes is investigated, while mutual exclusion among sequentially related distributed processes has been ignored. Our example here is a simple case of the latter mutual exclusion. Since parts have to be processed by M_1 first and then M_2, *unload_*M_1 and *unload_*M_2 can be mutually exclusive, and a sequential relation exists between two events. This book investigates both kinds of mutual exclusion in the Petri net context and applies the concepts and theoretical results to Petri net modeling and design of distributed resource-sharing systems.

Non-determinism: Two kinds of non-determinism may be observed. The first kind results from uncertain events' occurrence. For example, when there is a conflict as discussed above, either of two events can occur randomly, i.e., the robot can unload M_1 or unload M_2 without any certainty as to which will occur. Also, any two events which are ready to occur can take place in a random order. For example, when M_1 and M_2 are idle and there are parts available for their processing, either machine can be loaded, and the order is unpredictable. The second kind of non-determinism results from small changes in process parameters. For example, processing times of a machine differ from time to time due to randomness. Therefore, it cannot be accurately predicted when an event, e.g., ***complete_a_job***, will occur. In order to fully explore all the non-determinism in such systems, stochastic timed models need to be used. However, this book will focus on the logical models of the systems. The second kind of non-determinism (due to randomness) will not be discussed in this book.

System Deadlock: Suppose that, in the above system, the buffer capacity is 1. If we initially have 3 parts, M_1 processes the first part, and the robot unloads this intermediate part to the buffer. Then M_1, and M_2 loads and begins to process this intermediate part. While this is occurring, M_1 loads a second part and processes it, and the robot unloads this intermediate part to the buffer. M_1 proceeds to load and process the third part. loads a second part and processes. The robot then grasps this part and tries to unload M_1, the system is deadlocked. This is because M_1 cannot be unloaded because the buffer is already full. M_2 cannot be unloaded either because the robot is not available.

To avoid such a catastrophic failure, one can adopt either of two approaches. The first approach is that we design a supervisory controller which may lead to a deadlock. When it is implemented in a real-time environment, if a deadlock results, the system detects and resolves it. It is obvious that using such a method may be very costly since it may be expensive to detect and resolve deadlocks in a fully automated factory or a distributed system. For the system above, if the deadlock is detected, the recovery methods can be, for example, to use another robot to unload M_2 so the buffer becomes idle again. An alternative recovery method is to find a spare place which can be used for the temporary placement of the part grasped by the robot, and to direct the robot to place the part so that it becomes available to load M_2. We can observe that such an approach may lead a system to chaos if, for example, no good deadlock resolution methods are available.

The second approach is to design a supervisory controller with desirable system properties, for example, freedom from deadlock. In the above example, such a supervisory controller can be designed such that when the buffer is full and Machine 2 is busy, the robot is not allowed to unload Machine 1. Several mechanisms exist to implement such a controller. One of the simple methods is to limit the number of jobs which is allowed to enter this system to two in a particular time interval. Thus, when buffer is full and Machine 2 is busy, no job can be allowed to enter the system. Therefore, the robot cannot unload Machine 1.

The preceding example is simple and the solutions are intuitive. However, for complex systems, systematic methods are necessary to design supervisory controllers which have the desirable system properties. This book will address such a design problem and develop a design methodology for discrete event systems.

Summarizing, the increasing use of computer-based controllers to execute supervisory tasks in systems makes the study of event-driven system theory very important.

The next section gives an overview of modeling and control of discrete-event systems using different tools. In particular, the section stresses Petri net theory as a tool for designing supervisory controllers and presents the advantages of using Petri nets.

1.2 Modeling and Control of Discrete Event Systems

Discrete event systems exhibit many characteristics as described in the last section. In order to capture those properties, several mechanisms have been proposed and developed for modeling such systems. Among them are state machines, Petri nets, communicating sequential processes, and finitely recursive processes. In order to conduct performance analysis of these kinds of systems, methods such as perturbation analysis, queuing network theory, and Markov processes have been formulated and applied. In order to design supervisory controllers, automaton-based models have been used for such event-driven systems. This section begins with the brief discussion of hierarchical control structures of practical discrete event systems, and proceeds with the brief review of current approaches to modeling and control of DES. The advantages of using Petri nets for the investigation of design theory are discussed. Reviews of performance models for such systems can be seen in [Viswanadham 88a; Al-Jaar 89].

1.2.1 Hierarchical Control Structures

For manufacturing systems and other distributed systems, hierarchical control structures have been discussed in the literature [Gershwin 86; Menon 88; Valette 85a]. In this work, the control structure of a real-time event-driven system is viewed as a two-level structure. The first level control is the supervisory control or event-related control [Brand 88] and the second level is the process control which may be further decomposed and is often implemented by closed-loop feedback continuous or discrete controllers.

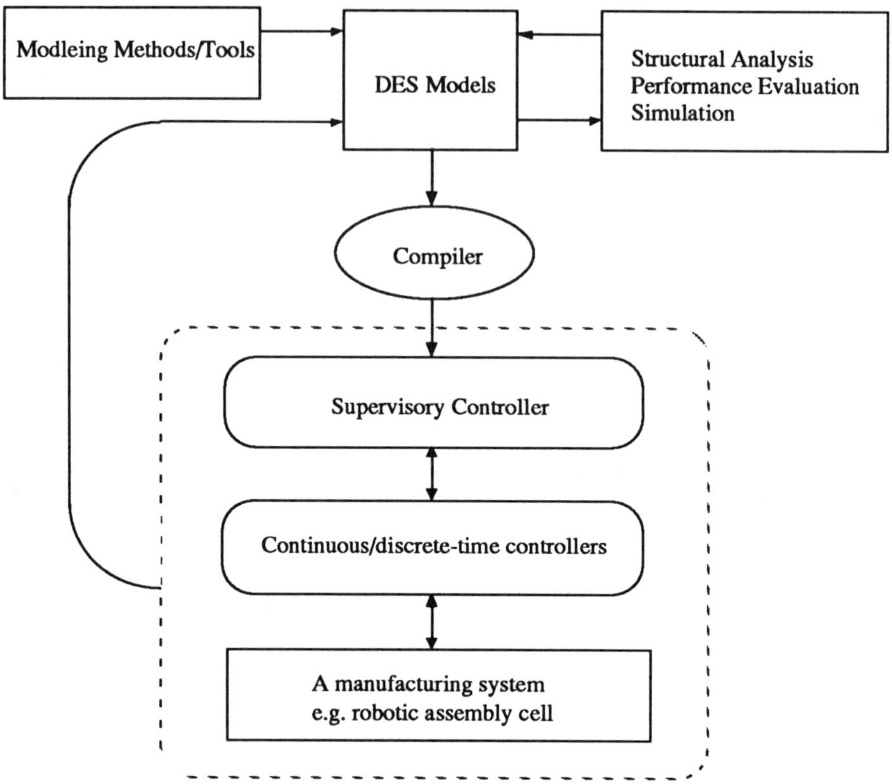

Figure 1.2: An architecture of the control of a real-world DES

Such a control structure can be observed through relating a manufacturing system to the architecture shown in Fig. 1.2. For example, when an event command "move the robot from point A to point B" is issued by a supervisory controller, the robot begins to move from A to B. The process "move" could be a continuous process which is controlled by some special controller. In a computerized environment, such a hierarchical control system can be implemented in two basic methods: centralized control and distributed control. In centralized control methods, a host computer is used to realize and synchronize all the event-related activities and low-level process controls. In distributed control methods, components of a system may be independent processor-based subsystems, and the communication between them becomes very important. A supervisory controller needs to be designed for coordination of asynchronous tasks among these distributed components to satisfy system requirements.

There already exist abundant theories for designing a continuous or discrete-time (synchronous) control system for low-level process control. However, the study of control theory and implementation of DES has just recently emerged as a focus of research. Therefore, this book hopefully contributes to the study of modeling and control of DES by addressing modeling methods and supervisory controller design approaches using Petri nets.

From Fig.1.2, one can see that DES models could be constructed using modeling approaches and tools for DES. Then such models can be used for supervisory controller design through a compiler or an algorithm. Other applications of these models include structural analysis, performance evaluation, and simulation for DES. One could also observe that a model for a DES may be directly converted into its supervisory controller. This holds true for Petri nets whose discussion follows.

1.2.2 Petri Net Model vs. Petri Net Controller

Petri net models for manufacturing systems have been widely used for the logical control validation, simulation, and performance analysis. They can also be implemented as the supervisory controllers of a system by imbedding the low-level processes into the places or transitions of its Petri net model. These low-level processes can be represented as software/hardware components [Naylor 87]. The controllers are represented as control code in the computer-based environment. Such a controller is called a Petri net controller which can be defined as a control-logic based on a marked Petri net for a discrete event system. The flow of tokens through places are regulated by the firing of transitions in a marked Petri net. It is this flow of tokens which defines the supervisory control actions.

Following Naylor and Volz [1987], software components are different from a module or subroutine in modular programming. They have three basic features:

1. A well-defined public interface;

2. An internal implementation that is inaccessible to the user; and

3. Both the visible part and the inaccessible implementation should be separately compilable from the program components that use them.

These features allow for preventing programming errors, extensive compiler error checking, and increasing program reliability, and make software components plug-compatible. The software/hardware component is the extension of the software component concept to include manufacturing components like a robot, or a machine.

The basic idea is that a "compiler" can be developed such that a Petri net model is transformed into, say, a C++ program to include the software/hardware components embedded in the places or transitions in the net. These components could be independently developed in advance and are used for low-level control of the system, for example, a continuous process: a robot moves an object from positions A to B. The resulting control programs and all software/hardware components will be integrated to execute the operation of an event-driven system, e.g., a manufacturing system.

Therefore, a Petri net model and its corresponding Petri net controller are equivalent in the sense that the latter can be converted from the former by using the "compiler" idea and cooperating low-level control processes. Therefore, a Petri net, a Petri net model, and a Petri net controller are considered as one unit as we later develop our Petri net synthesis method.

1.2.3 Petri Nets vs. Other Models in Modeling and Control

An event-driven system can be abstracted as a state machine in which the states change when events occur. The finite state machine or automaton model results when the total number of states in a system is finite. Finite state machines are well established as a fundamental model for computation and computing machines. However, when they are used to model DES in a straightforward manner, the exponential increase in the number of states makes it very difficult to implement complex DESs. Graphical representation is almost impossible and thus graphical visualization cannot be easily realized.

Other models have been recently developed for modeling and control of discrete event systems. Among them are formal language based supervisory control theory [Wonham 84, 87] and finitely recursive processes [Inan 88]. In supervisory control, the theory is elegant and is independent of the models used for applications. In most applications, each discrete event process is assumed to be modeled by an automaton or a state machine, and its behavior is completely described by the language generated by the automaton. All system requirements or specifications are also assumed to be specified as languages. Therefore, a design problem for supervisory controllers of discrete event systems can be stated as follows: find an automaton which is a supervisory controller such that the combined automaton for the controlled system generates the specified language. A supervisory controller controls a DES by enabling or disabling a controllable event. Many interesting theoretical results have been reported on controllability, observability and modular synthesis [Ramadge 83, 89; Wonham 84, 87]. However, the applicability to real-world distributed systems may be limited by the use of state machine representation. This approach encounters the state space explosion problem. Therefore, when a state machine is used to describe a complicated system, the design problem can easily become unmanageable. In addition, specifying the desirable language for a system is not easy.

Finitely recursive processes (FRP) have been proposed to describe discrete event systems or discrete event models (DEM) [Inan 88]. FRP are mainly based on Hoare's communicating sequential processes [Hoare 85]. In the FRP formulation, given a set of events, a process is defined as a triple which consists of three components: a set of traces which the process can execute, an event function, and a termination function. One of its important features is that each process can be described as a set of recursive equations. This implies that the description of a system can be implemented using equation forms. However, many problems remain open: for example, the use of such equations to design supervisory controllers for real-time systems.

Petri nets as a graphical tool provide a unified method for design of discrete event systems from hierarchical system descriptions to physical realizations. Compared with other models discussed, they have the following advantages [DiCesare 91; Ma 92; Martinez 86; Zhou 89a]:

1. Ease of modeling DES characteristics: concurrency, asynchronous and synchronous features, conflicts, mutual exclusion, precedence relations, non-determinism, and system deadlocks,

    Excellent visualization of system dependencies

    Focus on local information

    Top-down (stepwise refinement) design

    Bottom-up (modular composition) design.

2. Ability to generate supervisory control code directly from the graphical Petri net representation.

3. Ability to check the system for undesirable properties such as deadlock and instability and to validate code by mathematically based computer analysis - no time consuming simulations for many cases.

4. Performance analysis without simulation is possible for many systems. Production rates, resource utilization, reliability, and performability can be evaluated.

5. Discrete event simulation that can be driven from the model.

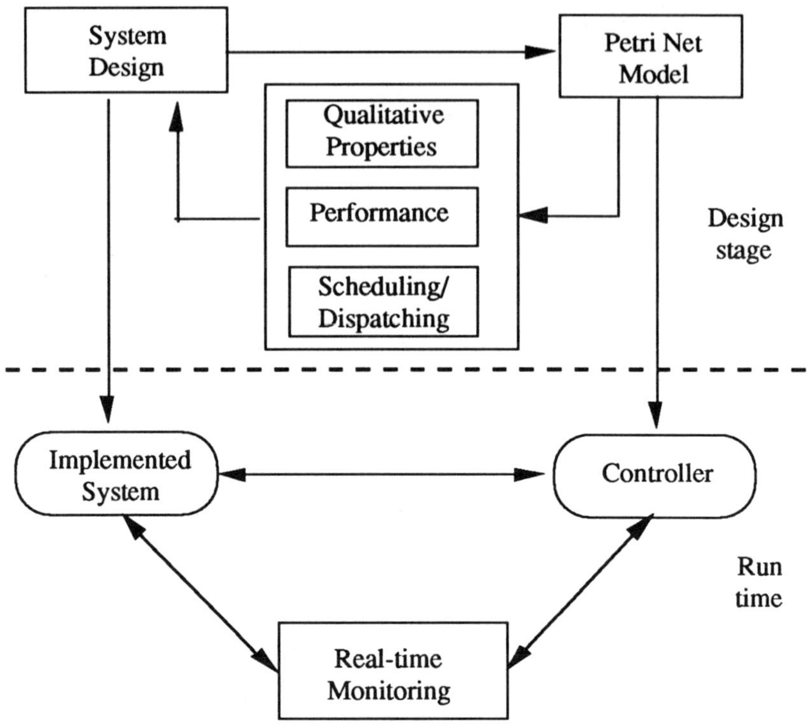

Figure 1.3: Petri nets as a single representation for modeling, analysis, and control

6. Status information that allows for real-time monitoring.

7. Usefulness for scheduling because the Petri net model contains the system precedence relations as well as constraints on discrete event performance.

As a single representation tool as shown in Figure 1.3, Petri nets can aid in modeling, analysis, validation, verification, simulation, scheduling, and performance evaluation at design stage. Once the system shows desirable behavior, the net can be converted into control and monitor operations at run time. Therefore, Petri nets can be regarded as a powerful mathematical and graphical tool for design of various discrete event systems.

1.3 Objectives of This Book

The goal of this book is to develop a design methodology for modeling and control of discrete event systems using Petri net theory. The application focus is on distributed systems such as automated manufacturing systems. The objectives are:

1. To present the definitions of Petri nets and properties related to manufacturing applications. The conventional methodology for Petri net synthesis is given and illustrated for a manufacturing system;

2. To propose parallel and sequential mutual exclusions which are fundamental concepts for the design of real-time resource-sharing distributed systems;

3. To develop a methodology for Petri net model and controller synthesis for manufacturing systems. The method adopts top-down design ideas to decompose and refine the nets in a synthesis process first, and then adopts bottom-up ideas to stepwise model all resources. The method should guarantee that the resulting Petri nets have the desired qualitative properties.

4. To apply the proposed methodologies to the modeling and control of distributed systems, e.g. an automated manufacturing system, a robotic assembly workstation, and a flexible manufacturing system (FMS). A resulting Petri net model is converted into a Petri net controller whose real-time implementation is illustrated for an FMS.

5. To develop and extend the Petri net controller concept for error recovery in automated manufacturing systems; when an error is recovered through the appropriate augmentation of Petri net controllers, the expected characteristics achieved by the original controllers will be preserved.

Chapter 2

Petri Nets and Manufacturing Systems

The purpose of this chapter is to define Petri nets and to introduce Petri net modeling in the context of manufacturing systems. The chapter begins with the formal definition of Petri nets and the properties of interest for manufacturing. This is followed by an example illustrating most of these concepts. Next, a methodology for modeling manufacturing systems with an example of its application is given. The example motivates the need for synthesis methods. Places of Petri nets are classified and additional concepts and assumptions are discussed to facilitate the presentation of the mutual exclusion theory developed later.

2.1 Petri Net Definitions and Properties

A *marked Petri net* $Z = (P, T, I, O, m_0)$ where:

P is a set of *places*, graphically represented by circles;

T is a set of *transitions*, graphically represented by bars, with $P \cup T \neq \emptyset$ and $P \cap T = \emptyset$;

$I : P \times T \rightarrow \{0, 1\}$ is the *input function* that specifies the arcs directed from places to transitions;

$O : P \times T \rightarrow \{0, 1\}$ is the *output function* that specifies the arcs directed from transitions to places;

$m : P \rightarrow N$ is a marking whose ith component is the number of tokens, graphically represented by dots, in the ith place. m_0 is an initial marking where $N = \{0, 1, 2, ...\}$.

The behavior of a Petri is determined by exercising *enabling* and *firing* rules:

A transition $t \in T$ is enabled iff $m(p) > 0$ when $I(p,t) = 1, \ \forall p \in P$.

An enabled transition t may fire at marking m', yielding the new marking $m(p_i) = m'(p_i) + O(p_i,t) - I(p_i,t)$ for $i = 1, 2, ..., |P|$. The marking m is said to be *reachable* from m'.

The definition of the *reachability set* and some net properties are as follows:

The *reachability set* is the set of all markings reachable from m_0 and is denoted by $R(Z, m_0)$.

A place $p \in P$ is *k-bounded* iff $\exists k \in N, \ni m(p) \leq k, \forall m \in R(Z, m_0)$.

Z is k-bounded iff p is k-bounded, $\forall p \in Z$. Z is *safe* iff it is 1-bounded.

Z is *live* iff $\exists$ a fireable sequence g, $\ni m[g>$ enables t, $\forall t \in T$ and $m \in R(Z, m_0)$.

Z is *reversible* iff $m_0 \in R(Z, m)$, $\forall m \in R(Z, m_0)$.

The significance of boundedness, liveness, and reversibility of Petri nets in manufacturing is discussed in the next section. In addition, the following concepts are useful in understanding the material presented in this book.

A Petri net is *consistent* iff $\exists m_0$ and a firing sequence g from m_0 to m_0 such that every transition occurs at least once in g.

Consistency of a Petri net implies that there is a firing sequence which brings a marking to itself by firing all transitions in the net. In manufacturing, this means that cyclic activities exist. However, consistency does not necessarily imply reversibility, whereas reversibility implies consistency. For example, sequence $t_1 t_2 t_3$ can bring m_0 to m_0 in a Petri net which consists of t_1, t_2, and t_3 only. Thus, the net is consistent. However, sequence $t_2 t_2$ may lead the system to a deadlock; then the net is not reversible.

Z is *conservative* iff $\exists$ a weighting vector $w > 0$, such that

$$\sum_{p \in P} w(p)m(p) = K \in N, \forall m \in R(Z, m_0).$$

A marked Petri net is called strictly conservative if $w = (1, 1, ..., 1)^\tau$. If a net is strictly conservative, then the sum of all tokens will remain unchanged for all reachable markings. Such Petri nets allow models for systems with a constant number of carts, fixtures, pallets, or jobs, e.g., a closed queueing system or a transportation system. Conservativeness implies boundedness.

Z is *persistent* iff $\forall m \in R(Z, m_0)$, an enabled transition can be disabled only by its own firing.

Persistence of a net precludes the existence of any conflict in a net. Hence, in shared resource allocation a persistent net model implies that there is no conflict among processes. Non-persistence of a net may imply some degree of unfairness in allocating resources in a manufacturing environment.

Several versions of synchronic distance definitions are available [Goltz 82, 86; Suzuki 83b; Wu 83; Yuan 86]. For safe and reversible Petri nets, all these definitions are unified as the following.

Synchronic distance,

$$\sigma(t, t') = \max_{g \in \Sigma(Z)} \{\sharp(g, t) - \sharp(g, t')\}$$

where $\Sigma(Z)$ represents all possible firing sequences in a safe and reversible Petri net Z given any binary initial marking.

$\sigma(t, t')$ is a measure of synchronic dependency between two transitions t and t'. If $\sigma(t, t') = \infty$, this assumes that unfairness exists since one transition can fire infinitely more than the other. The synchronic distances may also be used to obtain quantitative information about the dynamic behavior of a system without the introduction of a time notation, and can be viewed as a local clock or local controller in a distributed system.

An example will be used to show these concepts in the next section.

2.2 Modeling Manufacturing With Petri Nets

A straightforward view of manufacturing is a set of activities which interact with a set of resources and which result in a product. The activities are the manufacturing processes including machining, materials handling, and information processing, which must occur for production. The resources are the personnel, machines, raw materials and so on, which are required to perform these activities. The product process plan specifies the activities and resources in great detail. This includes the precedence relations between activities, i.e., some activities must occur before others, otherwise they could occur concurrently given the necessary resources. The process plan may specify several alternate resources for any activity and give some priority to which resource is preferred.

Petri nets as defined in graphical and mathematical form have no physical meaning. For use in applications, the modeler must provide the domain interpretation by the labeling of places and transitions. Generally in Petri net modeling, places (circles) represent conditions and transitions (bars) represent events. In our approach for modeling manufacturing systems with Petri nets, the following interpretations for places, transitions, and tokens are employed. These have been adopted by many researchers [Sifakis 77; Beck 86; Zhou 89a, 89b; Valvanis 90].

1. A place represents a resource status or an operation; when it represents the former, the initial number of tokens can be either a constant, e.g., the number of machines assuming the plant is fixed, or a variable, e.g., the number of jobs or pallets within the system.

2. If a place represents a resource status, one or more tokens in the place indicate that the resource is available and no token indicates that it is not available. If a place represents an operation, a token in it shows that an operation is being executed and no token shows that it is not being performed.

3. A transition represents either start or completion of an event or operation process.

The following example is presented to illustrate the definitions and the interpretation.

Example 2.1: Consider a transfer station as illustrated in Fig. 2.1. It consists of an AGV system and two workstations, and transfers two kinds of parts (F and G) produced by the two workstations using the AGV. The AGV transfers a part from either Workstation 1 or Workstation 2. Once it starts, it cannot be interrupted. As soon as the transfer is over, another part is made available and the AGV randomly transfers a part again.

A Petri net model of the above system is shown in Fig. 2.2 (a). The places and transitions have the following interpretation. See also Table 2.1. The places may represent resource availability or activity status as illustrated in Fig. 2.2. For example, places p_1, p_2 and p_5 in Fig. 2.2 model resource

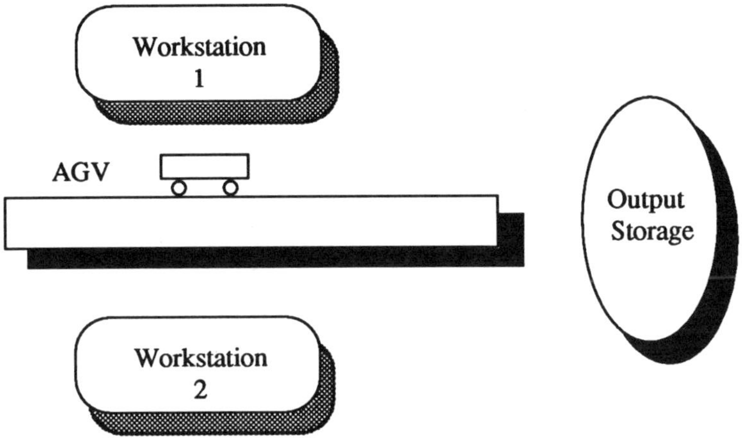

Figure 2.1: A part transfer station

Table 2.1: Interpretation of places and transitions

Places	*Transitions*
p_1: An F-part on WS_1 waiting for transfer	t_1: Transfer of an F-part starts
p_2: A G-part on WS_2 waiting for transfer	t_2: Transfer of a G-part starts
p_3: The F-part being transferred	t_3: Transfer of an F-part completes
p_4: The G-part being transferred	t_4: Transfer of a G-part completes
p_5: The AGV system ready to transfer a part	

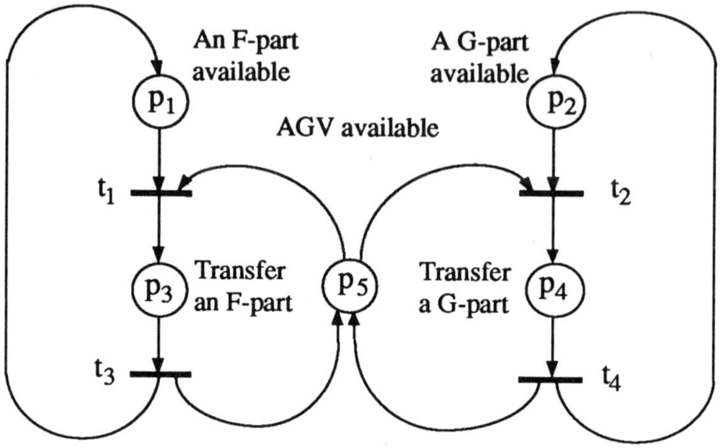

(a) A Petri net structure

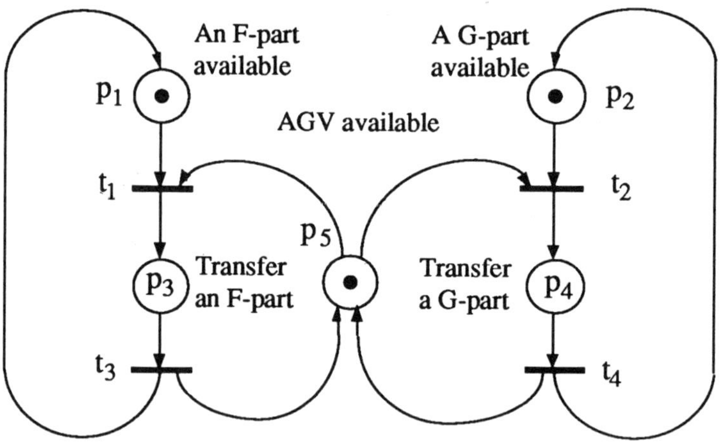

(b) Initial marking of the Petri net

Figure 2.2: A Petri net model for a part transfer station

availability status with p_1 indicating that an F-part is available when the place is marked by a token and not available when p_1 is not marked. Similarly p_2, when marked, indicates the availability of a G-part, the number of tokens representing the number of parts, and when unmarked, indicates no parts are available. Place p_5, when marked, indicates the availability of the AGV to transfer a part. The marking of places p_1 and p_5 represents the conditions that must jointly be true before the AGV transfer activity for an F-part can occur. The same is true for p_2 and p_5 with regard to the preconditions for the AGV transfer activity for a G-part. Places p_3 and p_4 model activity status, specifically p_3 being marked indicates the occurrence of the AGV F-part transfer activity and p_4 indicates a similar status for the AGV G-part transfer activity. Transitions t_1 and t_3, represent the starting and stopping events for the F-part transfer activity and t_2 and t_4 the same for the G-part.

For this Petri net, $P = \{p_1, p_2, p_3, p_4, p_5\}, T = \{t_1, t_2, t_3, t_4\}$.

The input function

$$I = \begin{pmatrix} 1 & 0 & 0 & 0 \\ 0 & 1 & 0 & 0 \\ 0 & 0 & 1 & 0 \\ 0 & 0 & 0 & 1 \\ 1 & 1 & 0 & 0 \end{pmatrix}$$

The output function

$$O = \begin{pmatrix} 0 & 0 & 1 & 0 \\ 0 & 0 & 0 & 1 \\ 1 & 0 & 0 & 0 \\ 0 & 1 & 0 & 0 \\ 0 & 0 & 1 & 1 \end{pmatrix}$$

A marked Petri net is given as shown in Fig. 2.2(b) with the initial marking:

$$m_0 = (1, 1, 0, 0, 1)^{\tau}$$

where the superscript τ denotes the transpose operation.

In this marking, both transitions t_1 and t_2 are enabled. If t_1 fires, the marking $m_1 = (0, 1, 1, 0, 0)^{\tau}$ is reached. If t_2 fires, the marking $m_2 = (1, 0, 0, 1, 0)^{\tau}$ is reached. At either m_1 or m_2, only one transition is enabled, t_3 or t_4. Firing either of them leads the net to its initial marking.

Based on the definitions of safeness, liveness, and reversibility, this Petri net is safe, live, and reversible by the definitions. Reversibility implies consistency. This is confirmed by the fact that the firing sequence $t_1 t_3 t_2 t_4$ brings $(1, 1, 0, 0, 1)^T$ to itself for the net structure shown in Fig. 2.2(a).

The net is not strictly conservative since

$$\sum_{p \in P} m_0(p) = 3 \neq 2 = \sum_{p \in P} m_1(p).$$

However, with $w = (1,1,2,2,1)^T$, the net is conservative.

The net is not persistent since the firing of t_1 can disable the enabled transition t_2. It is obvious that the sequence $t_1 t_3$ can fire infinitely without firing any other transitions. Hence, the synchronic distance between t_1 and t_2 is

$$\sigma(t_1, t_2) = \infty.$$

Similarly, we have

$$\sigma(t_2, t_1) = \infty$$

and

$$\sigma(t_1, t_3) = 1$$

implying that t_1 cannot fire more than once without t_3 firing once.

The infinite distance between t_1 and t_2 implies that the net could always fire t_1 without firing t_2. This means that the shared resource, i.e., the robot, may always be used to transfer F-parts in the example. Therefore, in this system the non-persistence or the infinite synchronic distance comments on the existence of unfairness.

2.2.1 Modeling Methods

Modeling methodology is key to the application of Petri nets to manufacturing systems. We start with one approach to modeling. Other approaches are possible and some will be used later in this book. As discussed earlier, the specification for a manufacturing system can be considered as a set of resources, a set of operations, and their precedence relationships. The following method [DiCesare 91] uses this specification to create a Petri net model of the system.

1. Identify the activities and resources required for the production of one item of each product.

2. Order activities by the precedence relations as given in the process plans.

3. For each activity in order: create and label a place to represent the status of that activity; add a transition (start activity) with an output arc(s) to the place(s); add a transition (stop activity) with an input arc(s) from the activity place(s). In general, the stop transition for one activity will be the same as the start transition for the next activity.

When the net is executed, a token in an activity place will indicate the activity is taking place. Multiple tokens will indicate the activity occurring in multiplicity, for example, in a buffer place, two tokens might represent two parts being stored at the same time. The firing of the start transition represents starting the activity or process and the firing of the stop transition represents the completion of the activity and may also represent the start of the next activity.

4. For each activity in order: If such a place has not been already created, create and label a place for each resource which must be available to start the activity. Connect all appropriate resource availability places with arcs such that each inputs to the starting transition for the activity. Create output arcs to connect the stop transition following the activity to any resource places representing resources which become available (are released) upon completion of the activity.

5. Specify the initial marking for the system.

To illustrate the modeling methodology, take the simple manufacturing system as shown in Fig. 1.1. As introduced earlier, it consists of two machining stations, M_1 and M_2, a shared robot, R, for unloading and a buffer, B, for storage of intermediate parts. The process plan for any part is straightforward: It must be machined by M_1, then M_2. Parts enter the system and at machining station M_1 each is automatically fixtured to a pallet and loaded into the machine. After processing, the robot unloads the intermediate part from M_1 into the buffer B. At machining station M_2 intermediate parts are automatically loaded into M_2 and processed. When M_2 finishes processing a part, the robot R unloads the final product, defixtures it and returns the fixture to M_1. It is assumed that input parts are always available to be fixtured and that the finished products are removed.

Following the methodology:

1. The activities required are machining (pallet fixturing, loading, and processing), storage (buffering) and unloading. The resources are M_1, M_2, R, B, pallets and parts.

2. The order of activities is as follows:

   M_1P: M_1 loads, fixtures and processes a palleted raw part.

   RU_1: R unloads an intermediate part to the buffer.

   BS: B stores an intermediate part.

   M_2P: M_2 loads and processes an intermediate part.

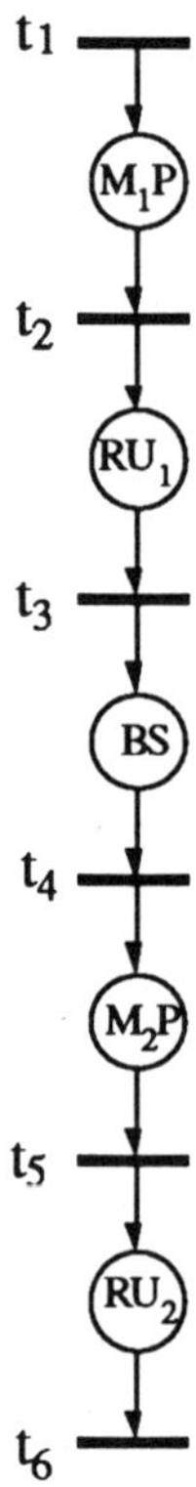

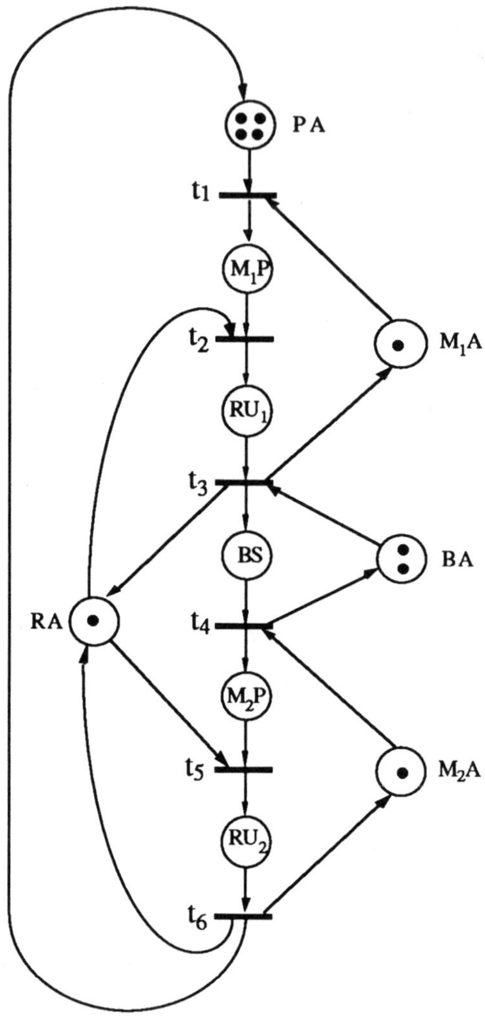

Figure 2.3: (a) Petri net graph of the activity sequence for the manufacturing example, (b) Petri net with the resource places added and connected and the initial marking

RU_2. R unloads a final product from M_2, defixtures and returns the pallet.

3. As shown in Fig. 2.3(a), places M_1P, RU_1, BS, M_2P, and RU_2 are created to model the activity sequence for one part. Transition t_1 models the start of activity M_1P; t_2 models the stop of activity M_1P and the start of activity RU_1; t_3, the stop of RU_1 and the start of the storage activity BS; t_4, the stop of BS and the start of the M_2P activity. Similarly, t_5 models the stop of activity M_2P and the start of RU_2 and t_6, the stop of RU_2.

4. For activity M_1P we require a pallet and machine M_1 be available. Fig. 2.3(b) shows the Petri net model with place PA representing pallets available, and place M_1A representing machine 1 availability. Each has an input arc to transition t_1. Next consider t_2 as the stop transition for activity M_1P. When the machining activity is complete, no resources can be released since the machine must be unloaded before becoming available again and the pallet will be used throughout manufacture.

   Considering t_2 as the start of the robot unloading machine 1 activity, RU_1, the only additional resource required is the robot R. This is modeled by the place RA representing robot available and the input arc from RA to t_2. With t_3 as the stop transition of activity RU_1 the release of M_1 can be modeled as an output arc from t_3 to M_1A and the release of R as an output arc from t_3 to RA.

   Next consider the buffer activity, BS. In order to store intermediate parts in the buffer, empty spaces in the buffer must exist. This is modeled by place BA where the number of tokens indicate the number of empty spaces and the arc from BA to t_3 the start transition of BS. The completion of the buffer activity for a part and the release of an empty space is represented by transition t_4 and the arc from t_4 to BA.

   The activities and resources for M_2P and RU_2 are modeled in the same way as for M_1P and RU_1 except for the output arc from t_6 to place PA. This represents the return of an empty pallet after M_2 is unloaded and the product is defixtured.

5. The initial marking is formulated for system startup. There are four pallets available for parts as indicated by the tokens in PA in Fig. 2.3(b); both machining stations and the robot are available as shown by the single tokens in M_1A, M_2A, and RA, respectively. There are two empty slots in the buffer for the storage of intermediate parts. Note that in this initial marking only transition t_1 is enabled.

It is interesting to observe the behavior of the system modeled in this manner. If the sequence t_1, t_2, t_3, t_4, t_1, t_2, t_3, t_1, t_2, t_3, t_1, t_2 is fired the system will be deadlocked. Of course with such a simple system, it is easy to identify the possibility of deadlock and prevent it either by changing the structure of the net or the initial marking of the net. This example provides motivation for the rest of this book. The goal is to find systematic synthesis procedures that will preserve systems properties such as liveness, boundedness and reversibility.

2.2.2 Net Properties and Manufacturing

One of the major advantages of Petri nets is the ability to analyze them for properties related to manufacturing control. These properties include boundedness or safeness, liveness, and reversibility. Their significance to manufacturing systems is stated as follows:

1. Boundedness or safeness implies the absence of capacity overflows. For instance, there may be storage buffers or queues which have a finite capacity. The boundedness of a place such as a buffer or queue insures that there will be no overflow. Safeness is the special case of 1-boundedness. As an example, safeness of an operation place guarantees that there is no attempt to request execution of an ongoing process. Safeness of a resource place indicates the availability of only a single resource and is often used to guarantee the safeness of some related operation places.

   To illustrate, place M_1A in Fig. 2.3 represents Machine 1 availability and its initial marking guarantees that places M_1P and RU_1 are safe, i.e., will never have more than one token, which means that the machine will never have more than one part at a time. Similarly, place BA's marking indicates the number of buffer spaces available and its initial marking guarantees that place BS is bounded, i.e., the buffer will never have more than two parts at any time.

2. Liveness implies the absence of deadlocks. This property guarantees that a system can successfully produce. Moreover, it insures that all modeled processes can occur.

   For the previous example, the initial marking may cause a deadlock as discussed earlier. However, if the initial marking is changed to have three tokens in place PA, then the system is live.

3. Reversibility implies the cyclic behavior of a system. This also has implications for error recovery in the manufacturing context. It means the system can be initialized from any reachable state.

2.2.3 Classification of Places

To review with regard to the method presented in this book, statuses, including those of operations and resources, are modeled as places in Petri nets. An operation can be divided into several sub-operations, and an operation place can be replaced by a subnet in a Petri net model. A token in an operation place implies that the operation in that place is being executed. Each operation place should be safe in the Petri net model because the same operation can never be executed more than once at the same time in the same machine. In this work, operation places are called *A places.*

One or more tokens in a resource place means that the corresponding resource is available, the number of tokens indicating the quantity. As we discussed earlier, resources can be further divided into two classes. The first represents resources whose number is fixed by the plant design, such as robots, machines, and conveyors. These resource places are called *B-places* in this work. The second class of resources has its number varied, for example, pallets, fixtures, and parts or jobs to be processed. Each operation generally needs both kinds of resources. The second class of resources plays a special role in the resource-sharing environment. In this book they are called *C-places.* Their number needs to be determined such that the system can be neither deadlocked nor starved. With Petri net models this is done by finding the appropriate number of initial tokens for these resource places.

Example 2.2: For the system discussed in Example 2.1 and the Petri net shown in Fig. 2.2, the places can be divided into the three groups as discussed.

1. A-places:
$$P_A = \{p_3, p_4\}.$$
2. B-places:
$$P_B = \{p_5\}.$$
3. C-places:
$$P_C = \{p_1, p_2\}.$$

Example 2.3: For the system discussed in the previous sections and the Petri net shown in Fig. 2.3, the places can be divided into the following groups.

1. A-places:
$$P_A = \{M_1P, RU_1, BS, M_2P, RU_2\}.$$

2. B-places:

$$P_B = \{RA, M_1A, BA, M_2A\}.$$

3. C-places:

$$P_C = \{PA\}.$$

The distinction among the resource places are not mandatory for a particular system. For example, one may classify place BA as a C-place since the buffer size is a variable in some manufacturing applications. This fact shows that the modeling or design process of a physical system can heavily depend upon the designer's choice and the system needs.

2.2.4 Other Related Definitions and Assumptions

In addition to Petri net definitions and properties discussed before, the following concepts and assumptions are useful for the presentation of the material in this book.

Given $Z = (P, T, I, O, m_0)$, ***a node*** is either a place in P or a transition in T. ***An elementary path*** is a sequence of nodes: $x_1x_2\cdots x_n, n \geq 1$, such that $\exists$ an arc (x_i, x_{i+1}) for $i \in N_{n-1}$ if $n > 1$, and $x_i = x_j$ implies that $i = j$, $\forall i, j \in N_n$. It is denoted by $EP(x_1, x_n)$ if $n > 1$. ***An elementary circuit*** is $x_1x_2...x_n, n > 1$ such that $x_i = x_j, 1 \leq i < j \leq n$, implies that $i = 1$, $j = n$. It is denoted by $EC(x_1)$. If $P = P_A \cup P_B \cup P_C$, an ***A-path*** between x and y is defined as an elementary path between x and y, each of whose nodes except x and y is either an A-place or a transition.

In the literature, the above defined elementary path is also called a simple elementary path [Berge 62].

For convenience, $EP(x_1, x_n)$ is used as a set of nodes, i.e., $\{x_1, x_2, ..., x_n\}$. $EC(x)$ is also treated as a set. Therefore, $x \in EP(x_1, x_n)$ means that x is a node on the elementary path between x_1 and x_n. Given $x, y \in P \cup T, EP(x, y)$ may represent any elementary path or the set of all elementary paths between x and y in particular contexts and may be used as a set. If there is no such path between x and y, then we say $EP(x, y) = \emptyset$. Also, $EP(x, Y) = \{x' : \exists y \in Y, EP(x, y), \ni x' \in EP(x, y)\}$ if $Y \subset P \cup T$, i.e., $EP(x, Y)$ is the set of all nodes on the elementary paths between node x and set Y.

An A-path starting and ending with transitions contains only initially non-marked places. In the manufacturing context, it may represent a series of operations which may be required to finish some types of jobs.

Example 2.4: For the Petri net depicted in Fig. 2.2(a), we have the following elementary paths and circuits:

$$EP(p_1, t_3) = p_1t_1p_3t_3 \text{ or } \{p_1, t_1, p_3, t_3\}$$

$$EP(t_1, t_2) = t_1p_3t_3p_5t_2 \text{ or } \{t_1, p_3, t_3, p_5, t_2\}$$

$$\begin{aligned} EP(\{p_1, p_2\}, t_3) &= EP(p_1, t_3) \cup EP(p_2, t_3) \\ &= \{p_1, t_1, p_3, t_3\} \cup \{p_2, t_2, p_4, t_4, p_5, t_1, p_3, t_3\} \\ &= \{p_1, t_1, p_3, t_3, p_2, t_2, p_4, t_4, p_5\}. \end{aligned}$$

There are two elementary circuits starting and ending with t_1, i.e.,

$$EC(t_1) = \{t_1p_3t_3p_1t_1, t_1p_3t_3p_5t_1\} \text{ or } \{t_1, p_3, t_3, p_1, p_5\}.$$

Also, there are two elementary circuits around p_5, i.e.,

$$EC(p_5) = \{p_5t_1p_3t_3p_5, p_5t_2p_4t_4p_5\} \text{ or, } \{p_5, t_1, p_3, t_3, t_2, p_4, t_4\}.$$

Suppose that the places are classified into A, B, and C-places as discussed in Example 2.3. According to the definition of an A-path, $EP(p_1, t_3)$ is also an A-path. There is no A-path between t_1 and t_2, but there is an elementary path.

B-places and C-places model the availability of resources which are necessary to start some operations. Otherwise, a designer could certainly discard such a place from the net. Therefore, the following assumption is made for the classification of places.

Assumption 2.1 *Given a Petri net $Z = (P_A \cup P_B \cup P_C, T, I, O, m_0)$, where P_A, P_B, and P_C are the sets of A-places, B-places, and C-places respectively, and are mutually disjoint, the output transitions of p' in $P_B \cup P_C$ are never enabled, $\forall p' \in P_B \cup P_C, m \in R(Z, m_0)$ if initial marking $m_0(p') = 0$ and*

$$m_0(p) = \begin{cases} 0 & \text{if } p \in P_A; \\ \geq 1 & \text{if } p \in P_B \cup P_C - \{p'\}. \end{cases}$$

This assumption also implies that any resources in a place p cannot be converted to ones in another place $p', \forall p, p' \in P_B \cup P_C$. The assumption is necessary to formulate mutual exclusion concepts discussed in the next two chapters.

We introduce the following example to comment on Assumption 2.1.

Example 2.5: Consider the transfer station described in Fig. 2.1 and its Petri net model in Fig. 2.2(a). Here, we have operation place set $P_A = \{p_3, p_4\}$, fixed resource place set $P_B = \{p_5\}$, and variable resource place set $P_C = \{p_1, p_2\}$, as we indicated in Example 2.1. P_B and P_C obviously satisfy the above assumption. Note that the net is bounded, live, and reversible whenever there are tokens initially in places p_1, p_2, and p_5. It may be noted that if p_1 and p_5 are two B-places in the Petri net in Fig. 2.2(a), then p_3 would not be a B-place since p_3 does not satisfy Assumption 2.1. In fact, as above, p_3 represents an operation and is an A-place.

In the following discussion, it is assumed that

$$m_0(p) = \begin{cases} 0 & \text{if } p \in P_A \\ constant(\geq 1) & \text{if } p \in P_B \\ variable(\geq 1) & \text{if } p \in P_C \end{cases}$$

Only C-places are allowed to have their number of initial tokens varied.

We say an initial marking m_0 is *feasible* if and only if the marked Petri net $Z = (P_A \cup P_B \cup P_C, T, I, O, m_0)$ is bounded, live, and reversible. Therefore, finding a feasible initial marking of a Petri net structure with $(P_A \cup P_B \cup P_C, T, I, O)$ is equivalent to finding $m_0(p) \in N^+$, $p \in P_C$ such that $(P_A \cup P_B \cup P_C, T, I, O, m_0)$ is bounded, live, and reversible where $N^+ = \{1, 2, 3, \ldots\}$.

The following discussion assumes that the tokens of any two or more resource places, i.e., B- and C-places can flow into a common place through only the same transition. This is true for manufacturing systems because each resource place is used to model the availability of machines, robots, raw materials, fixtures, or pallets, and if two tokens represent two different things in different places, they cannot be mixed in the same place which may represent some operation. However, the tokens are allowed to represent different contents going to an identical place through a common transition, i.e., through synchronization. Such synchronization may start an operation, for example, fixturing the raw material and machining a part. Choice places with several output transitions are also permitted. Therefore, the following assumption is formulated.

Assumption 2.2 *Given a marked Petri net $Z, p_1, p_2 \in P_B \cup P_C, t \in T$, if $p_1 \neq p_2$, and if $\exists EP(p_1, t)$ and $EP(p_2, t)$ which are A-paths, then their first intersection occurs at a transition.*

Note that the first intersection of two paths $a_1a_2..a_n$ and $b_1b_2...b_m$ occurs at $a_i = b_j, i \leq n, j \leq m$ if $a_u \neq b_v, 0 < u < i$ and $0 < v < j$. In addition, one can easily find examples to illustrate that Assumption 2.1 does not imply Assumption 2.2 and vice versa.

For two C-places p_1 and p_2 in the Petri net discussed in Fig. 2.2 (b), there is a transition, say t_3, such that $EP(p_1, t_3)$ and $EP(p_2, t_3)$ exist. Their first intersection occurs at transition t_1. Note that $EP(p_2, t_3)$ is not an A-path. Obviously, p_1, p_2, and p_5 satisfy Assumption 2.2 since their A-paths either intersects at the resource place, i.e., $p_5t_1p_3t_3$ and $p_5t_2p_4t_4$, or never intersect, i.e., $p_1t_1p_3t_3$ and $p_2t_2p_4t_4$.

2.3 Summary

This chapter has defined the Petri net terminologies necessary to state the more advanced concepts and theory for Petri net synthesis. It has also discussed the conventional synthesis methods which may lead to a Petri net which contains undesirable properties. The experience motivates the research on synthesis methodologies presented in this book. In summarizing, a general problem for modeling a manufacturing system using a Petri net can be stated as follows:

> *Given the specifications of a manufacturing system, model the system as a Petri net such that its structure and initial markings make it bounded, live, and reversible.*

Then, the resulting Petri net model can be translated to supervisory control code for actual execution of a manufacturing system. The related techniques and implementation details will be explored in this book.

Chapter 3

Parallel Mutual Exclusions

3.1 Introduction

Applications of Petri net theory have been extended to various aspects of manufacturing since its introduction to this domain in the 1970's. Among them are specification, validation, performance analysis, and simulation for automated manufacturing systems [Al-Jaar 90a; Desrochers 89; DiCesare 91; Murata 89; Silva 90; Zhou 90a]. One can also directly convert Petri net definitions into control code for supervisory control of manufacturing systems. Only after correct Petri net models are constructed for systems do all these applications follow. Therefore, the systematic modeling of a system is very important. Traditionally, one obtains a Petri net model using trial and error methods, and then analyzes it for properties like boundedness and liveness. Since in a moderately sized manufacturing system the complexity of design at the implementation level of detail easily becomes unreasonable, a challenging problem arises: that of systematically constructing a Petri net model such that it has the desirable properties for a manufacturing system.

One major difficulty of the problem lies in the method of dealing with the shared resources in the system [Jeng 92; Koh 91; Zhou 90a, b]. These resources can be robots, material movement systems, or machining centers. The existence of shared resources is a principal reason why a system may contain deadlocks. For example, the simple manufacturing system introduced in Chapter 1 may be deadlocked due to the existence of the shared robot. One can be easily convinced that it would never be deadlocked without the shared robot no matter how many parts are present and processed.

One objective of this book is to address the synthesis problem for a system with shared resources and present a rigorous approach. This chapter presents parallel mutual exclusion and its role in synthesis. Two major objectives of this chapter are:

1. To formulate the parallel mutual exclusion (PME) resource-sharing concept in the context of Petri net theory; and

2. To study the conditions under which Petri nets containing such structures are live, bounded, and reversible.

To capture characteristics of a system with a resource shared by independent processes, we define parallel mutual exclusion (PME) [Zhou 91a]. The next section gives a basic definition of parallel mutual exclusion and an example. Properties and theorems about the qualitative properties of Petri nets containing parallel mutual exclusion structures are developed and proved in Section 3.3. An application example on the modeling of a two-robot-two-assembly-line system is given in Section 3.4. The parallel mutual exclusion concept is generalized in Section 3.5.

3.2 Definition and Example

By considering the execution of the system model for Example 2.1, we can observe the following properties:

1. When the shared resource place p_5 is not considered, the system contains two independent processes, each of which can be repeated infinitely; and

2. When the shared resource is considered, two processes cannot be executed simultaneously. However, both are ready to be executed at the initial state, and either can be executed after the on-going process is complete.

The observations and study of the above example have motivated us to define the following parallel mutual exclusion concept. In the definition, the ith one of k processes is modeled as a pair of transition (t_{ai}, t_{bi}), and the shared resource is modeled as a place p_E. In Example 2.1, $k = 2$, (t_1, t_3) and (t_2, t_4) model two processes, and $p_E = p_5$.

Definition 3.1 Given a marked Petri net

$$Z = (P, T, I, O, m_0),\ \ P = P_A \cup P_B \cup P_C,$$

a *k-parallel mutual exclusion* is defined as k-PME$=(p_E, D)$ such that

1. $p_E \in P_B$ with $m_0(p_E) = 1$, D is a set of transition pairs or

$$D = (t_{a1}, t_{b1}), (t_{a2}, t_{b2}), ..., (t_{ak}, t_{bk}),\ k \geq 1,$$

satisfying:

   (a) $t_{ai}, t_{bi} \in T, t_{ai} \neq t_{bj}, t_{ai} \neq t_{aj}$, and $t_{bi} \neq t_{bj}$ when $i \neq j, \forall i, j \in N_k$;
   (b) $I(p_E, t_{ai}) = O(p_E, t_{bi}) = 1, I(p_E, t_{bi}) = O(p_E, t_{ai}) = 0$ when $t_{bi} \neq t_{ai}, \forall i \in N_k; I(p_E, t_u) = O(p_E, t_u) = 0$ when $t_u \notin T_a \cup T_b$, where $T_a = \{t_{ai}, i \in N_k\}$, and $T_b = \{t_{bi}, i \in N_k\}$;
   (c) $\forall i \in N_k, p \in P_C$, if $p \in EC(t_{ai}), t_{bi} \in EC(t_{ai})$;
   (d) $\forall EC(t_{ai})$, if $EC(t_{ai}) \cap (P_B \cup P_C) = \{p_E\}, t_{bi} \in EC(t_{ai})$; and
   (e) if $t \in EP(t_{ai}, t_{bi}), t$ is on an A-path $EP(t_{ai}, t_{bi})$;

2. $P_B \cup P_C$ satisfies Assumptions 2.1 and 2.2 and there is no A-path between t and t', $\forall t \in T_b, t' \in T_a$;

3. $\exists m_0$, and a sequence of fireable transitions g that contains no transition in $T_a, \ni m_0[g\rangle$ enables $t, \forall t \in T_a$; and

4. $\forall m_0$, if t_{aj} fires at $m \in R(Z, m_0)$, then $\forall t$, if $EP(t_{aj}, t) \neq \emptyset$ and $t_{bj} \notin EP(t_{aj}, t), t$ can be enabled and $\forall$ fireable g_j containing no $t_{bj}, \exists h_j, \ni m[t_{aj} g_j h_j\rangle$ enables t_{bj}; if $m(p_E) = 1$ and t_u is enabled, then $t_u \notin T_b$;

A k-PME characterizes a resource shared by k processes. Each process, which may be very complicated, is abstracted and expressed as a pair of transitions between which there may be a complicated subnet. For convenience, D is considered to be a set of all the transitions in $T_a \cup T_b$. Therefore, $t_u \notin D$ means $t_u \notin T_a \cup T_b$.

Conditions 1 and 2 in this definition define the structure of mutual exclusion in a Petri net, called *a PME structure.* Conditions 3 and 4 define the behavior that is possessed by a parallel mutual exclusion and is dependent on the classification of places.

Detailed explanation of Conditions 1 to 4 is as follows:

Condition 1:

(a) All transitions in D should be different except t_{ai} and t_{bi}. Thus t_{ai} can be equal to t_{bi}. Practically, it is possible and an example may be referred to in a Petri net model for a two-process system [Agerwala 79] . Another example using self-loops to model a single shared resource can be seen in [Magott 85, 87] . However, we exclude such a special case in further discussion since the existence of a pair (t_{ai}, t_{bi}) does not affect the properties of the Petri nets when $t_{ai} = t_{bi}$.

(b) There is one input arc from p_E to each t_{ai} and one output arc from t_{bi} to p_E for $1 \leq i \leq k$ but no other arcs related to p_E.

(c) Any elementary path between t_{ai} and a C-place has to contain t_{bi} for $1 \leq i \leq k$. It together with (b) implies that after t_{ai} fires and consumes the shared resource as well as resources originating from the variable resources, i.e., C-places, these resources will be released only through firing of t_{bi}.

(d) Any elementary circuit having t_{ai}, the shared resource place p_E, but neither the other B-places nor C-places, must include t_{bi} for $1 \leq i \leq k$; and

(e) Each transition on an elementary path between t_{ai} and t_{bi} should be on one A-path between t_{ai} and t_{bi} for $1 \leq i \leq k$. It guarantees that the concurrent groups of operation processes in the system be successive.

Condition 2: It requires that all resources be well modeled in the sense that all resource places satisfy Assumption 2.1 which says that if a resource place is not initially marked, its output transition is never marked. Assumption 2.1 also implies that any resources whose status are modeled as different places cannot be converted into each other. Moreover, Assumption 2.2 ensures that tokens for different resources such as robots and machines cannot be mixed together. Condition 2 also requires that no sequential relations exist among the k processes.

Condition 3: It guarantees equal opportunity for each process to compete for and acquire the resource.

Condition 4: It guarantees the proper use and release of the shared resource from the process. It also assures that any transition whose firing needs t_{aj} can be enabled after t_{aj} fires where $1 \leq j \leq k$.

Example 3.1: In the Petri net shown in Fig. 2.2, $(p_5, \{(t_1, t_3), (t_2, t_4)\})$ is a 2-PME when $P_A = \{p_3, p_4\}, P_B = \{p_5\}$, and $P_C = \{p_1, p_2\}$. However, when $P_A = \{p_3, p_4\}, P_B = \{p_5, p_2\}$, and $P_C = \{p_1\}$ and $m_0(p_2) = 0, (p_5, (t_1, t_3), (t_2, t_4))$ is not a 2-PME from the above definition because it cannot satisfy Condition 3. If two arcs (p_5, t_2) and (t_4, p_5) are deleted, $(p_5, \{(t_1, t_3)\})$ constitutes a 1-PME, which is a trivial case.

1-parallel mutual exclusion $(p_E, \{(t_a, t_b)\})$ represents the availability of a single resource plus two related transitions as shown in the Fig. 3.1(a). Figure 3.1(b) can be a simple case for such a 1-PME which is a case of a single circuit for marked graphs [Murata 80a]. For ordinary Petri nets, if a 1-PME results

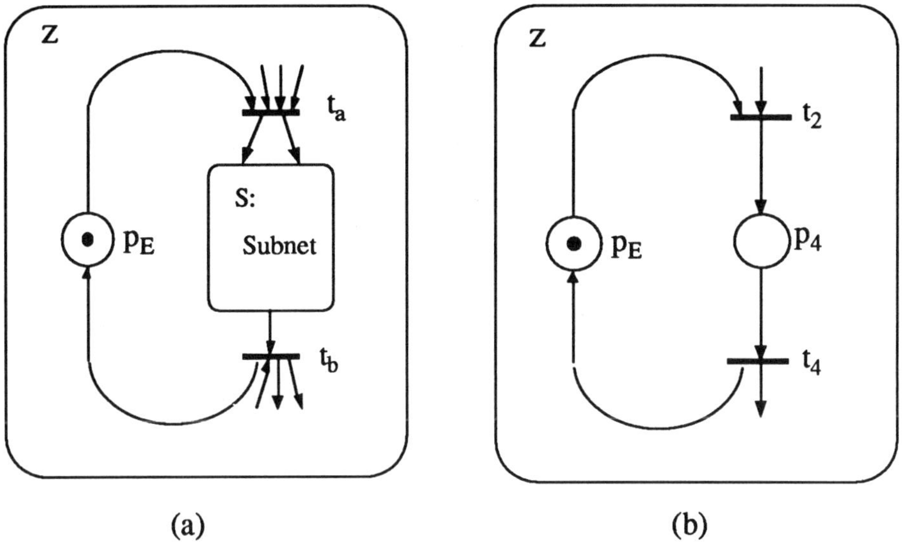

Figure 3.1: 1-parallel mutual exclusion examples, (a) a general case and (b) a simple case

from an addition of a B-place plus two arcs to two transitions, then the addition is later shown to preserve the boundedness or safeness, liveness, and reversibility of the original net. This will become the basis for the deductive proof for the preservation of properties of a Petri net having a k-PME.

Theorems in the next section claim that the boundedness, liveness, and reversibility are preserved for the Petri net which results from the addition of a k-parallel mutual exclusion to a bounded, live, and reversible Petri net. A generalization of a k-PME can be made by extending (t_{ai}, t_{bi}) in D to (T_{ai}, T_{bi}) where both T_{ai} and T_{bi} are subsets of T and will be discussed in Section 3.5.

3.3 Development of Theoretical Results

This section presents the main theoretical results about the preservation of the properties of a Petri net containing a parallel mutual exclusion. A property of a k-PME is first described as a lemma.

Lemma 3.1 *Assume that there is a k-PME* (p_E, D) *where* $D = \{(t_{ai}, t_{bi}), i \in N_k\}$ *in a Petri net* $Z' = (P_A \cup (P_B \cup \{p_E\}) \cup P_C, T', I', O', m'_0)$. *If* $\exists$ *a sequence of transitions* f *containing no* $t_{bi}, \ni t_{ai} f t_{bi}$ *fires from* $m \in R(Z', m'_0)$, *then* f *contains no* t_{aj}, $\forall j \in N_k$, *and* $i \in N_k$.

Proof: From Condition 4 in the definition of a k-PME, the firing of t_{ai} implies that the unique token in p_E is removed and therefore, no transition t with $I(p_E,t) = 1$ can be enabled without the firing of the transition $t_{bi}, \forall i \in N_k$. This shows that t_{aj} cannot fire,$\forall j \in N_k$. Therefore, f contains no $t_{aj}, \forall j \in N_k$. QED

This lemma states that if the shared resource is occupied by a process, then it cannot be used by any other process until it is released from that process.

Lemma 3.2 *p_E is a safe place in Z'.*

Proof: According to Condition 4 in Definition 3.1, no transition in T_b can be enabled when $m(p_E) = 1$. This ensures that $m(p_E) \leq 1$ for any marking m in Z'. Therefore, p_E is a safe place. QED

The following theorems reveal the fact that if a system contains several independent processes and has some desirable qualitative properties, then when these processes need to compete for a resource with equal opportunity, and release it after that process acquires and uses it, the system still exhibits the desirable properties.

The following assumptions are made for Theorems 3.1 to 3.3 which deal with a Petri net which has parallel mutual exclusion:

1. Given $Z' = (P', T', I', O', m'_0)$ with $P' = P_A \cup (P_B \cup \{p_E\}) \cup P_C$ and $T' = T, Z = (P_A \cup P_B \cup P_C, T, I, O, m_0)$ is a subnet of Z'. In other words, $\forall p \in P$, $t \in T$, $I(p,t) = I'(p,t)$, $O(p,t) = O'(p,t)$, and $m_0(p) = m'_0(p)$.

2. (p_E, D) is a k-PME of Z' where $D = \{(t_{ai}, t_{bi}), i \in N_k\}, k > 0$.

Theorem 3.1 *Z' is bounded (safe) if Z is bounded (safe).*

Proof: Let Z^k denote the marked Petri net with a k-PME (p_E, D^k), and $R(Z^k, m'_0)$ the reachability set of Z^k from the initial marking m'_0. $m(p_E)$ is assumed to be the last element of any marking $m \in R(Z^k, m'_0)$. Also, write

$$m = \begin{pmatrix} \mu \\ m(p_E) \end{pmatrix}$$

as

$$(\ \mu \quad m(p_E)\ )$$

to save space. A sequence of fireable transitions is denoted by f, g, or h, and $m[f>$ means that f is applied to the marking m.

Define $Q(Z^k, m_0') = \{\mu : (\mu, m(p_E)) \in R(Z^k, m_0')\}$ which is a set of vectors. Its element μ has the same components as m except that it does not contain $m(p_E)$.

If $m \in (Z^k, m_0'), \exists h, \ni m_0'[h> m$, i.e., $(m_0, 1)[h> (\mu, m_0(p_E))$. Since h can be applied to Z, $m_0[h> \mu$. Therefore, $m \in R(Z, m_0)$. Since $R(Z, m_0)$ is bounded according to the assumption, so is μ. $m(p_E)$ is safe using Lemma 3.2. $R(Z_k, m_0')$ is bounded. Therefore, Z_k is bounded if Z is bounded. Similarly, the conclusion can be shown for safeness. QED

Theorem 3.2 *Z' is live if Z is live.*

Proof: The same notation is used as the proof of Theorem 3.1. The proof is conducted by induction on k as follows.

First, when $k = 1$, this case deals with the addition of a single place p_E which is initially marked with a single token, and has only two arcs to two different transitions t_{a1} and t_{b1}, i.e., $m_0(p_E) = 1$, and $I(p_E, t_{a1}) = O(p_E, t_{b1}) = 1$.$\forall t \in T, m \in R(Z^1, m_0')$, we need to show that there is a sequence of transitions whose firing enables t. If $m(p_E) = 1$, then we already have $m = (\mu, 1)$. Otherwise, suppose that $m(p_E) = 0$, which implies that the number of firings of t_{a1} is one more than that of t_{b1}. Since $(p_E, \{(t_{a1}, t_{b1})\})$ is a 1-PME, there is a sequence of transitions which enables t_{b1}. Suppose that firing this sequence plus t_{b1} leads the marking to m' with $m'(p_E) = 1$. Let $m' = (\mu, 1)$.

$\mu \in R(Z, m_0)$ according to the proof of Theorem 3.1. For μ and t in Z, there are two cases: $EP(t_{a1}, t) \neq \emptyset$ or $EP(t_{a1}, t) = \emptyset$.

For $EP(t_{a1}, t) \neq \emptyset$, since Z is live according to the assumption, $\exists g$ which contains no t_{a1} in $Z, \ni \mu[g> \mu'$ which enables t_{a1}. The reason why g contains no t_{a1} is that before t_{a1} fire from m, it must be enabled. This g can fire in Z^1 obviously, which implies that $m'[g> m''$ enables t_{a1}. At m'', t can be enabled according to Condition 4 in Definition 3.1 of PME if $t_{b1} \not\in EP(t_{a1}, t)$. Otherwise, since $t_{b1} \in EP(t_{a1}, t)$, A-path $EP(t_{b1}, t) \neq \emptyset$. In $Z, \exists h, \ni \mu[g> \mu''$ which enables t. Such h can be obviously arranged so that each firing of t_{a1} is followed by a t_{b1}. Therefore, h is fireable and enables t in Z^1. This proves the first case.

For the case where $EP(t_{a1}, t) = \emptyset$, the liveness of Z implies that $\exists g, \ni \mu[g> \mu'$ which enables t. g can obviously be chosen to contain no t_{a1}. Therefore, it is fireable in Z^1 and makes t enabled. This accomplishes the proof for the case when $k = 1$.

Second, suppose that for $k \leq n$, the conclusions are true. The following shows that they are so for $k = n + 1$.

Suppose $(p_E, D^{(n+1)}) = (p_E, D^{(n)} \cup (t_a, t_b))$ is a $(n + 1)$-PME. We need to show that t can be enabled, $\forall t \in T, m \in R(Z^{n+1}, m_0')$. There are two possibilities:

1. $m \in R(Z^n, m_0')$, i.e., m also belongs to $R(Z^n, m_0')$; and

2. If $m = (\mu, m(p_E)) \notin R(Z^n, m_0')$, then $(\mu, 1) \in R(Z^n, m_0')$.

By the induction assumption, $\exists g$ in $Z^n, \ni m[g>$ or $(\mu, 1)[g>$ enables t. The following proof focuses on the first case. A similar proof can be constructed for the second possibility. There are three subcases, A, B, and C and the proof for each is as follows:

Case A: If g contains neither t_a, t_b, nor the transitions on any elementary paths between t_a and t_b, then it is fireable in Z^{n+1}. Therefore, $m[g>$ also enables t in Z^{n+1}.

Case B: If $m(p_E) = 1$ and g contains t_a once or several times, then we need to show that $\exists g', \ni m[g'>$ enables t in Z^{n+1}. Since $(p_E, D^{(n+1)})$ is and $(n+1)$-PME, g can be arranged into one of the following two forms without changing the resulting marking:

$$g_1 t_a f_1 t_b g_2 t_a f_2 t_b \cdots t_a f_v t_b g_{v+1}, v \geq 1$$

or

$$g_1 t_a f_1 t_b g_2 t_a f_2 t_b \cdots g_v t_a f_v, v \geq 1$$

where f_i and g_i contain neither t_a nor t_b. g_i contains no transition in $EP(t_a, t_b)$.

B.1 Consider the first form

$$g = g_1 t_a f_1 t_b g_2 t_a f_2 t_b \cdots t_a f_v t_b g_{v+1}, v \geq 1$$

Using Lemma 3.2 of n-PME in Z^n, g_i contains at most one element of $T_a = \{t_{ai}, i \in N_n\}$, say t_{aj} without the following of the transition t_{bj}. If there is, for example, $t_{aj} h_i t_{bj}$ in g_i, then $t_{aj} h_i t_{bj}$ can be applied in Z^{n+1} where h_i contains no transitions in $T_a, \forall i \in N_{v+1}$. Therefore, we assume that, for $i \in N_v$, $g_i = t_{aj} h_i$ and $f_i = r_i t_{bj} e_i$ without loss of generality, and $\exists w_i = g_i t_a f_i t_b = t_{aj} h_i t_a r_i t_{bj} e_i t_b$ in $g, \ni w_i$ cannot fire in Z^{n+1}. We will prove that $w_i' = t_{aj} h_i r_i t_{bj} t_a e_i t_b$ can fire, and the resulting marking is the same as that by firing w_i in Z^n.

a. w_i' is fireable in Z^n (proof by contradiction).

If w_i' cannot fire in Z^n, this implies that there is a transition, say t' in $r_i t_{bj}$ whose firing needs firing of both transitions t_{aj} and t_a. This would be true in Z^{n+1}. However, $(p_E, D^{(n+1)})$ is an $(n+1)$-PME and $D^{(n+1)} = D^{(n)} \cup \{(t_a, t_b)\}$ which precludes the existence of such a transition t'. This is a contradiction. Therefore, w_i' is fireable in Z^n.

b. $\eta(p)[w_i'> = \eta(p)[w_i>, \forall p \in P, \eta \in R(Z^n, m_0')$.

Following the firing rule of a marked Petri net, $\forall p \in P$, and $t, t' \in T$, if both tt' and $t't$ at $\psi \in R(Z^n, m_0')$ are fireable, then

$$\psi(p)[t't> = \psi(p) - I(p,t') + O(p,t') - I(p,t) + O(p,t)$$

and

$$\psi(p)[tt'> = \psi(p) - I(p,t) + O(p,t) - I(p,t') + O(p,t')$$

Thus

$$\psi(p)[t't> = \psi(p)[tt'>$$

Since both sequences w_i' and w_i are firable and consist of the same transitions, the conclusion holds true.

Similarly, we can show that

$$m(p)[t_a h_i t_{aj} r_i t_b e_i t_{bj}> = m(p)[t_a h_i r_i t_b t_{aj} e_i t_{bj}>, \forall p \in P.$$

Using the above technique, g can be converted into g' which can fire and $g'(m)$ enables t in Z^{n+1}.

B.2 Consider the second form

$$g = g_1 t_a f_1 t_b g_2 t_a f_2 t_b \cdots g_v t_a f_v, v \geq 1$$

Using the above techniques, g' can be obtained from g, which can fire and $m[g'>$ enables t in Z^{n+1}.

Case C: If $m(p_E) = 0$ and g contains t_a and transitions on the paths between t_a and t_b once or several times, similarly, we can show that g' can be found such that $m[g'>$ enables t in Z^{n+1}. In conclusion, any transition t in Z^{n+1} is live. QED

Theorem 3.3 *Z' is reversible if Z is reversible.*

Proof: The proof is similar to the proof of Theorem 3.2 and is also conducted by induction on k.

$\forall m \in R(Z^{n+1}, m_0')$, we need to show that $\exists g$ in $Z^{n+1}, \ni m[g> m_0'$. Since either $m \in R(Z^n, m_0')$ or $(\mu, 1) \in R(Z^n, m_0'), \exists g$ in $Z^n, \ni$ either $m[g> m_0'$ or $(\mu, 1)[g> m0'$. Using similar reasoning as for the proof of the liveness of Z^{n+1}, the conclusion can be proved. QED

Remark 1. All these conditions are sufficient, but not necessary. For example, the first conclusion in Theorem 3.1 implies that Z' may be bounded or safe even if Z is unbounded. One example is shown in Fig. 3.2 where the Petri net in (a), i.e., Z, is unbounded because of the existence of the unbounded places p_4 and p_5. The Petri net in Fig. 3.2(b), i.e., Z' is obtained by adding a resource place p_6 and thus contains a 2-PME. This net is safe. It is easily seen that both nets are live and reversible. The liveness and reversibility of the net in Fig. 3.2 (b) can also be obtained from Theorems 3.2 and 3.3. A reader can construct Petri net Z and Z' such that Z is neither live nor reversible but Z' is live and reversible.

Remark 2. The theorems hold true for the multi-resource cases in the sense that the conditions in the theorems hold true for each newly added mutual exclusion structure in the net. In many cases, if one finds that any PME conditions are unsatisfied for a specific Petri net, then a deadlock can be easily found if it exists. An example of this will be shown in the next section.

Remark 3. The proof of the theorems suggests a way for Petri net synthesis with shared resources by stepwise addition of arcs to a shared resource. One can start with a 1-PME, then add a pair of arcs to 1-PME to form a 2-PME, ... until the shared resource is completely modeled. When more than one shared resource are included, one can add the mutual exclusion structures one by one in a systematic manner. The example of two robot two workstation assembly system in the next section will illustrate this synthesis procedure.

3.4 Application Example

This section will demonstrate the usefulness of the above theoretical results in Petri net modeling of a distributed system. The modeling technique introduced in Chapter 2 is used to model a two-robot-two-workstation assembly system. The system can be viewed as a manufacturing system version of the simplified Dining Philosopher's Problem [Dijkstra 68]. The resulting Petri net will be shown to have a possible deadlock. Then the parallel mutual exclusion concept and results will be used to model a deadlock-free system.

3.4.1 System Description

This robotic assembly system consists of two robot manipulators: R_1 and R_2 and two workstations: WS_1 and WS_2 as shown in Fig. 3.3. The specification of this system is as follows.

1. When either of WS_1 and WS_2 is ready to execute the assembly task, it requests its right robot and acquires it if it is available.

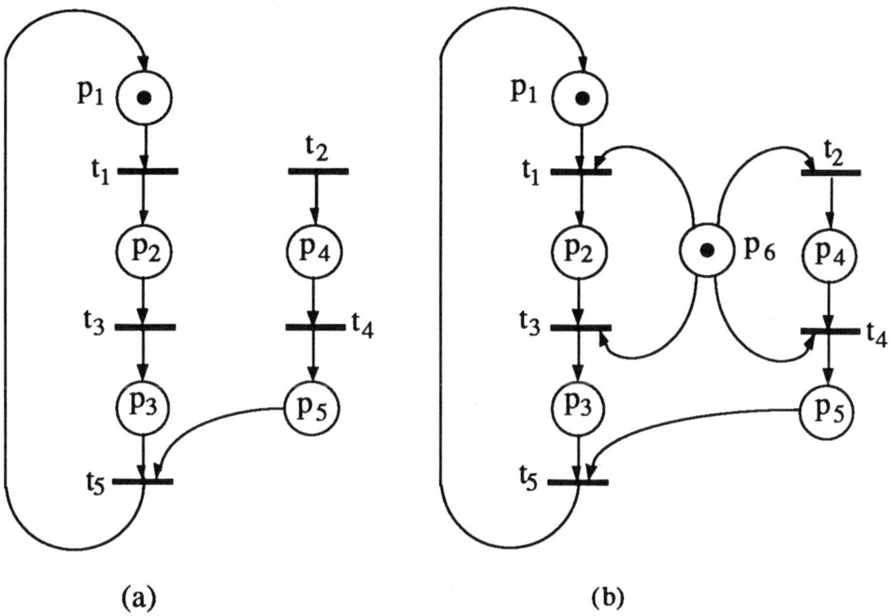

Figure 3.2: Two Petri nets which comment on Theorem 3.1, (a) Z with $m_0 = (1,0,0,0,0)^T$ (b) Z' with $m_0 = (1,0,0,0,0,1)^T$

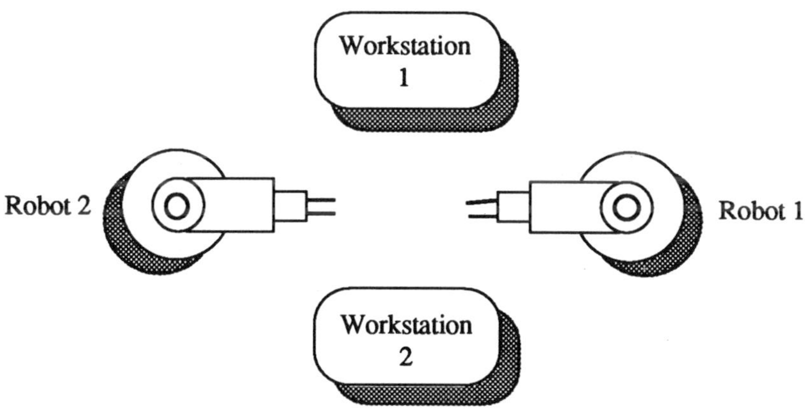

Figure 3.3: The layout of a two-robot and two-workstation assembly system

Table 3.1: List of system activities

WS_1	WS_2
Acquiring R_1	Acquiring R_2
Acquiring R_2	Acquiring R_1
Assembling	Assembling

2. After a workstation acquires its right robot, it requests its left robot and acquires it if it is available.

3. When a workstation starts an assembly task, it cannot be interrupted until it completes.

4. When WS_1 (WS_2) completes, it releases both robots.

3.4.2 Petri Net Modeling

For this system, its Petri net model can be easily designed using the methodology introduced in Chapter 2. Four steps are briefly discussed as follows:

1. The activities required are: acquiring its right robot, acquiring its left robot, and assembling for both workstations.

2. The order of the activities is shown in Table 3.1.

3. As shown in Fig. 3.4, the places p_2, p_3, and p_4 are created for WS_1 and represent the three activities. The same holds for p_6, p_7, and p_8 for WS_2.

4. Acquiring R_1 in p_2 requires the request of R_1 from WS_1 and robot R_1 be available. They are represented by place p_1 and p_9, respectively. Each has an input arc to transition t_1. t_1 represents the start of the operation in p_2. This is modeled by an arc (t_1, p_2). After R_1 is acquired, the next operation is to acquire R_2, which requires R_2. There is an input arc to t_2 from p_2 and one from p_1 which models the availability of R_2. After R_2 is acquired, WS_1 starts assembling. This is modeled by one input arc from p_3 to t_3 and one output arc from t_3 to p_4. After assembling, two robots are released. This is represented by two output arcs from t_4 to p_9 and p_{10}. In order to model repetitive request signals, an arc from t_4 to p_1 is also included. Similarly, we can model the other part for WS_2.

5. The initial marking is $(1,0,0,0,1,0,0,0,1,1)^T$ which means that initially there are two requests from WS_1 and WS_2, respectively and two

Table 3.2: Place and transition labels for the Petri net

Places	*Transitions*
p_1: WS_1 requests R_1	t_1: Start acquiring R_1 at WS_1
p_2: WS_1 acquires R_1	t_2: Start acquiring R_2 at WS_1
p_3: WS_1 acquires R_2	t_3: Start assembling at WS_1
p_4: Assembling at WS_1	t_4: Complete assembling at WS_1
p_5: WS_2 requests R_2	t_5: Start acquiring R_2 at WS_2
p_6: WS_2 acquires R_2	t_6: Start acquiring R_1 at WS_2
p_7: WS_2 acquires R_1	t_7: Start assembling at WS_2
p_8: Assembling at WS_2	t_8: Complete assembling at WS_2
p_9: Availability of R_1	
p_{10}: Availability of R_2	

robots are available. The explanation of all places and transitions are summarized in Table 3.2 and the net is shown in Fig. 3.4.

3.4.3 Petri Net Analysis and Modification

Two mutual exclusion structures (parallel mutual exclusion) of this net are $(p_9, \{(t_1, t_4), (t_6, t_8)\})$ and $(p_{10}, \{(t_2, t_4), (t_5, t_8)\})$. Each would form a 2-PME without the other in the net. However, neither of them is a 2-PME because neither satisfies Condition 4 in Definition 3.1. For example, consider in Fig. 3.4 $(p_9, (t_1, t_3), (t_5, t_6))$ when the marking is $m_0' = (1,0,0,0,1,0,0,0,1,1)^T$. Both transitions t_1 and t_5 are enabled. However, t_3 cannot fire after t_1 fires. In fact, the system can be deadlocked; neither process can proceed when the marking $(0,1,0,0,0,1,0,0,0,0)^T$ is reached after firing t_1 and t_4 or t_4 and t_1 from m_0'.

The method proposed and discussed in Chapter 5 suggests that designers should consider shared resources one by one. Thus Fig. 3.5 is obtained after robot R_1 is included. $(p_9, \{(t_1, t_4), (t_6, t_8)\})$ is easily verified to be a 2-PME.

As we mentioned before, when $(p_{10}, \{(t_2, t_4), (t_5, t_8)\})$ is added according to the system specification, it is not a 2-PME. Thus the given system specification is not correct for the design of a deadlock-free system. Let's modify the specification as follows: Workstation 1 requests both robots at the same time. In other words, it will start the assembling only when two robots are available and can then be reserved for Workstation 1. Thus we get the net shown in Fig. 3.6 after $(p_{10}, \{(t_1, t_4), (t_5, t_8)\})$ is constructed. Next we will show that it is a 2-PME.

By noting that $P_A = \{p_2, p_3, p_4, p_6, p_7, p_8\}, P_B = \{p_1, p_5, p_9, p_{10}\}$, and $P_C = \emptyset$, one can easily verify Conditions 1 and 2. For Condition 3, the initial marking has already enabled both t_1 and t_5. For condition 4, since

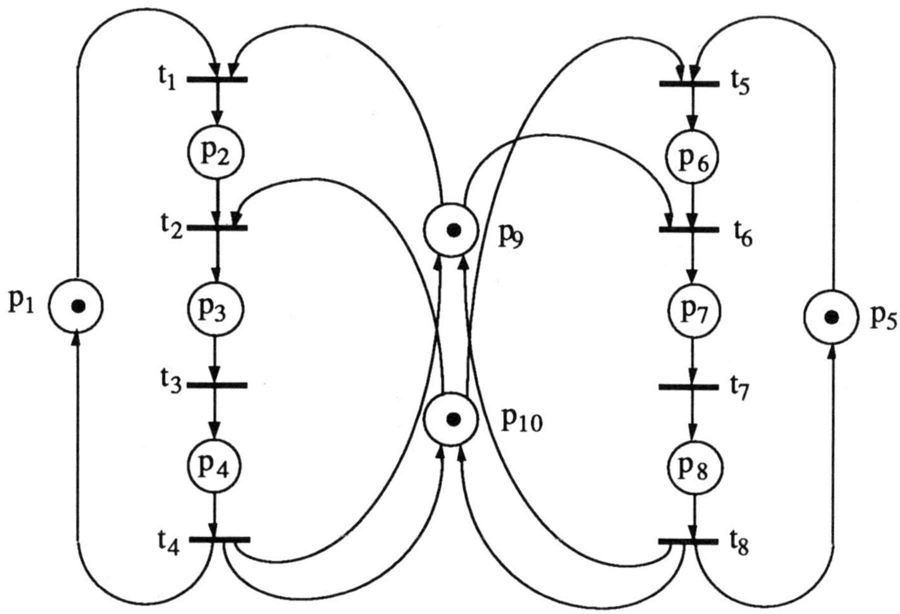

Figure 3.4: A Petri net model for the system

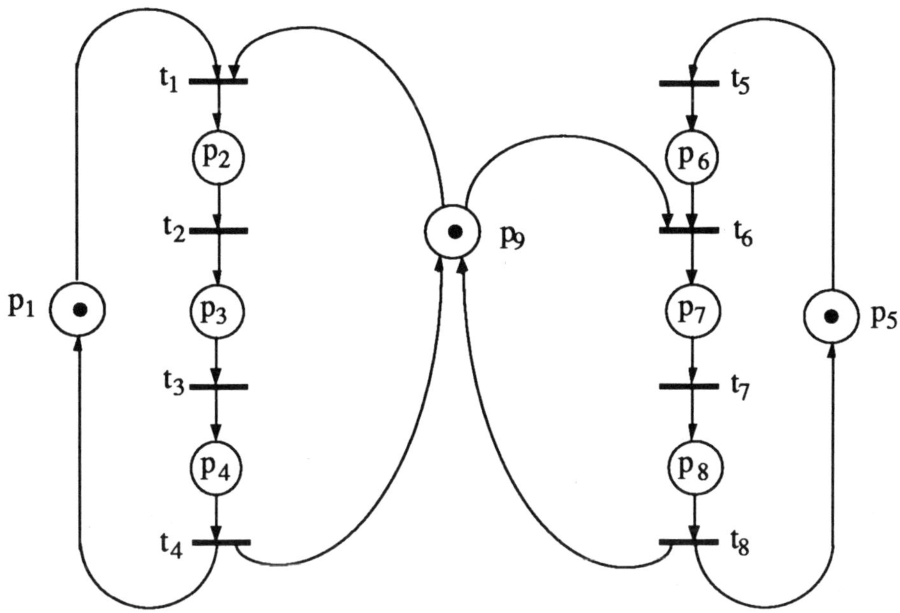

Figure 3.5: A Petri net model when only R_1 is considered

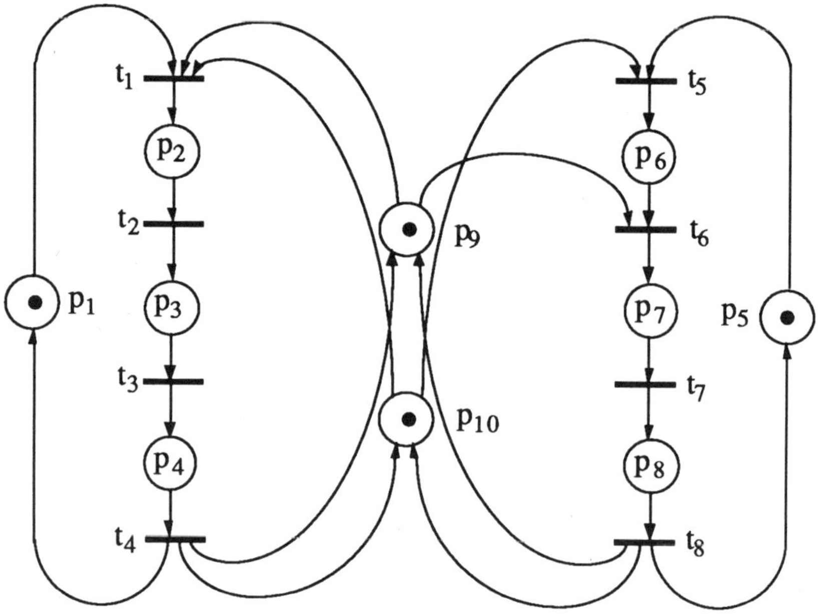

Figure 3.6: A Petri net with two 2-PMEs

$P_C = \emptyset$, we need to consider $m'_0 = (1,0,0,0,1,0,0,0,1,1)^T$ only. If t_1 fires, then t_2, t_3, and t_4 are the unique sequence which can fire. If t_5 fires, t_6, t_7, and t_8 are a unique firable sequence. Thus Condition 4 is true. Therefore, $(p_{10}, \{(t_1,t_4),(t_5,t_8)\})$ is a 2-PME.

Since the net in Fig. 3.5 is safe, live, and reversible, the net in Fig. 3.6 is also safe, live, and reversible according to Theorems 3.1-3.3.

Through the discussion of Petri net modeling of the preceding assembly system, some interesting observations can be made. Using this theory ensures the intermediate Petri nets to be deadlock-free through addition of a parallel mutual exclusion for robot R_1. However, when the second mutual exclusion structure is taken into account, it cannot constitute a parallel mutual exclusion. This actually implies a deadlock for the Petri net model shown in Fig. 3.4.

Two possible strategies follow: the first method is to add the last shared resource to form a parallel mutual exclusion such that the system is modified to have no deadlock. This implies that the system specification has to be changed as we did in the example. The second method is to build up the net model in the "correct" way although the previous parallel mutual exclusions may be degraded to non-parallel mutual exclusions. Then some previous deadlock avoidance methods [Datta 84; Hsieh 92] need to be used to prevent such built-in deadlocks. Other examples by using parallel mutual exclusion concepts can be found in [Zhou 91b, c].

In conclusion, the results reported give insight into the Petri net modeling of distributed systems which contain built-in deadlocks. One advantage of our method lies in the fact that the potential deadlocks may be detected during the modeling process. Furthermore, such early findings can greatly decrease the design cost for real-time resource-sharing distributed systems, especially when the system becomes very complicated.

3.5 Generalized Parallel Mutual Exclusion

3.5.1 Definition

Parallel mutual exclusions can be used to resolve a synthesis problem where the shared resource deals with no choice operations. These concepts may also be extended to a more general case where choice operations may exist. For example, a robot can be in charge of unloading between Machine 1 and two successive machines, say, Machines 2 and 3. When it acquires a processed part from Machine 1, the robot can load the part on either Machine 2 or Machine 3. However, the existing mutual exclusion structure cannot describe the preceding situation where the robot is a shared resource. Therefore, it is highly desirable to generalize these ideas, and this section describes this work.

Definition 3.2 Given a marked Petri net $Z' = (P', T', I', O', m_0), P' = P_A \cup P_B \cup P_C$, a k-generalized parallel mutual exclusion is defined as k-GPME=(p_E, D) such that

1. $p_E \in P_B$ with $m_0(p_E) = 1$, D is a set of the transition pairs or $D = (T_{a1}, T_{b1}), (T_{a2}, T_{b2}), ..., (T_{ak}, T_{bk}), k \geq 1$, satisfying:

   (a) $T_{ai}, T_{bi} \subset T, T_{ai} \cap T_{bj} = \emptyset, T_{ai} \cap T_{aj} = \emptyset$, and $T_{bi} \cap T_{bj} = \emptyset$ when $i \neq j, \forall i, j \in N_k$;

   (b) $I(p_E, t) = O(p_E, t') = 1, I(p_E, t') = O(p_E, t) = 0$ when $\forall t \in T_a, t' \in T_b$, and $I(p_E, t_u) = O(p_E, t_u) = 0$ when $t_u \notin T_a \cup T_b$, where $T_a = T_{a1} \cup T_{a2} \cup \cdots \cup T_{ak}$ and $T_b = T_{b1} \cup T_{b2} \cup \cdots \cup T_{bk}$;

   (c) $\forall t \in T_{ai}$ and $t' \in T_{bi}$, A-path $EP(t, t') \neq \emptyset$;

   (d) $\forall t \in T_{ai}, p \in P_C$, if $p \in EC(t)$, then $\exists t' \in T_{bi}, \ni t' \in EC(t)$;

   (e) $\forall t \in T_{ai}$, if $EC(t) \cap (P_B \cup P_C) = p_E$, then $\exists t' \in T_{bi}, \ni t' \in EC(t)$; and

   (f) if $t'' \in EP(t, t'), t$, then t'' is on an A-path $EP(t_{ai}$ where $t \in T_{ai}, t' \in T_b), i \in N_k$;

2. $P_B \cup P_C$ satisfies Assumptions 2.1 and 2.2 and there is no A-path between t and t', $\forall t \in T_b, t' \in T_a$;

3. $\exists m_0$, for a group of transitions $\{r_{a1}, r_{a2}, ..., r_{ak}\}$ where $r_{ai} \in T_{ai}, \exists g$ that contains no transition in $T_a - \{r_{ai}, i \in N_k\}, \ni m_0[g>$ enables $r_{ai}, i \in N_k$; and

4. $\forall m_0$, if $r \in T_{aj}$ fires at $m \in R(Z, m_0)$, then $\forall t$, if $EP(r, t) \neq \emptyset$ and $T_{bj} \cap EP(r, t) = \emptyset, t$ can be enabled and $\forall$ fireable g_j containing no transitions in $T_{bj}, \exists h_j$ and $r' \in T_{bj}, \ni m[rg_jh_j>$ enables t'; and if $m(p_E) = 1$ and t_u is enabled, then $t_u \notin T_b$;

Remarks similar to those made for PME (see Section 3.1) can be made for the above GPME definition. We also note two additional interesting facts:

1. When $|T_{aj}| = |T_{bj}| = 1$, $\forall i \in N_k$, then the above k-GPME is a k-PME as defined in Section 3.1.

2. When $|T_{aj}| \leq 2, |T_{bj}| \leq 2$, $\forall i \in N_k$, we can enumerate most cases as shown in Fig. 3.7. The in Fig. 3.7(a) and (b) cannot be used to constitute any GPMEs defined above. However, subnets in Fig. 3.7(c)-(f) can be used to constitute GPMEs.

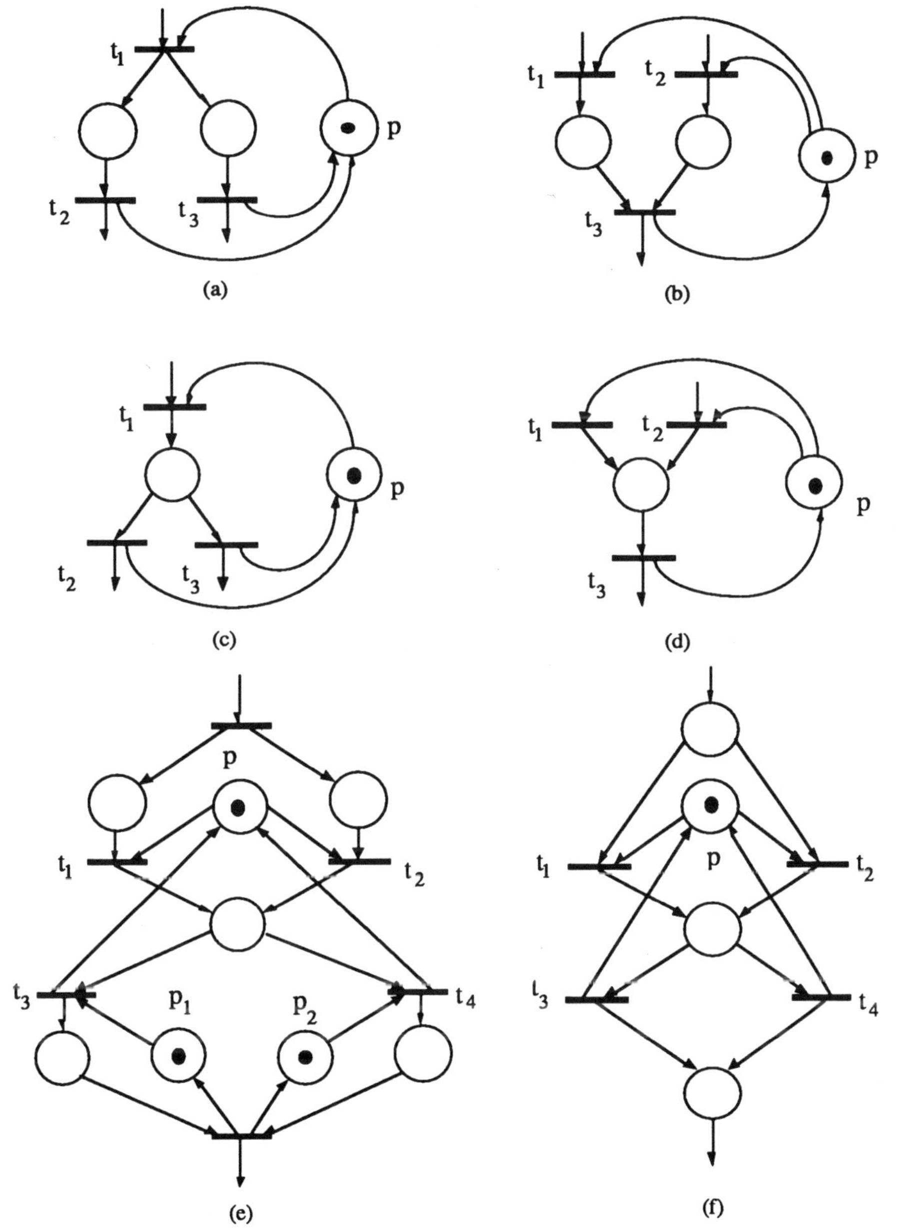

Figure 3.7: Sub-Petri nets where (a) and (b) cannot constitute a GPME and the other can

In Fig. 3.7(a), after t_1 fires, p will obtain two tokens by firing t_2 and t_3. However, this is contrary to the fourth condition in the above definition. When Fig. 3.7(b) is considered, firing t_3 requires the firing of both t_1 and t_2. This is impossible since only one token in p is available. However, (c) and (d) can be used to constitute 1-GPMEs, i.e., $(p, \{(t_1, \{t_2, t_3\})\})$, and $(p, \{(\{t_1, t_2\}, t_3)\})$, respectively.

Subnets shown in Fig. 3.7(e) and (f) can also be used as two examples of 1-GPMEs since one can verify that $(p, (\{t_1, t_2\}, \{t_3, t_4\}))$ in either net satisfies the conditions given in the above definition. In Fig. 3.7(e), places p_1 and p_2 are crucial to 1-GPME of the net. Without any of them, the structure may lead to a deadlock. For example, discarding p_1 can make the net partially deadlocked if the system is bounded. This is because we can let t_3 fire only whenever t_3 and t_4 are enabled. The boundedness guarantees that tokens will be exhausted in the common input place of t_3 and t_4; thus t_3, t_4, etc. cannot be enabled from this state.

3.5.2 Preservation of System Properties

Using the techniques to prove Theorems 3.1 to 3.3, we can prove the following theorem for a k-GPME. This theorem generalizes Theorems 3.1 to 3.3 for a Petri net containing a k-GPME.

Theorem 3.4 (Generalized parallel mutual exclusion) $Z = (P_A \cup P_B \cup P_C, T, I, O, m_0)$ *is a subnet of* $Z' = (P_A \cup (P_B \cup \{p_E\}) \cup P_C, T', I', O', m_0'), T' = T$, *and* $D = \{(T_{ai}, T_{bi}), i \in N_k\}$. *If* (p_E, D) *is a k-GPME, then*

1. *Z' is bounded (safe) if Z is bounded (safe);*

2. *Z' is live if Z is live; and*

3. *Z' is reversible if Z is reversible.*

As we have noted for PMEs, all the conditions in this theorem are sufficient, but not necessary. Other remarks for PMEs in Section 3.3 can be modified to comment on this theorem.

The case for a non-shared resource occurs when $k = 1$ in the definition of parallel mutual exclusion, and when $k = 1$ and $|T_{ai}| = |T_{bi}| = 1$ in the definition of generalized parallel mutual exclusion. It describes the following structure: $t, t' \in T$, one A-path $EP(t, t'), p \in P$, and $m_0(p) = 1$, such that $I(p, t) = 1, O(p, t') = 1$. In this case, the following theorem is derived as a special case of Theorem 3.4.

Corollary 3.1 (Addition of a non-shared resource place) *Suppose that* $Z = (P_A \cup P_B \cup P_C, T \cup \{t_a, t_b\}, I, O, m_0)$ *is a subnet of* $Z' = (P', T', I', O', m_0')$,

$P' = P_A \cup (P_B \cup \{p_E\}) \cup P_C, T' = T \cup \{t_a, t_b\}$. *If* $(p_E, (t_a, t_b))$ *is a* 1-*GPME, then*

1. *Z' is bounded (safe) if Z is bounded (safe);*
2. *Z' is live if Z is live; and*
3. *Z' is reversible if Z is reversible.*

This corollary is very useful when we synthesize the non-shared resource places in the design procedure for Petri nets in Chapter 5.

Example 3.2: A Petri net is shown in Fig. 3.8. In this Petri net, the places are classified as follows:

1. Operation place set: $P_A = \{p_2, p_3, p_4, p_5, p_9, p_{10}, p_{11}\}$,
2. Fixed resource place set: $P_B = \{p_6, p_7, p_{12}, p_E\}$, and
3. Variable resource place set: $P_C = \{p_1, p_8\}$.

Each B-place is initially marked as shown in Fig. 3.8. It can be easily verified that (p_E, D) is a 2-GPME, where $D = \{(t_1, \{t_2, t_3\}), (t_7, t_8)\}$ and $m_0 = (m_0(p_1), 0, 0, 0, 0, 1, 1, m_0(p_8), 0, 0, 0, 2, 1)^\tau$ whenever $m_0(p_1) > 0$, and $m_0(p_8) > 0$.

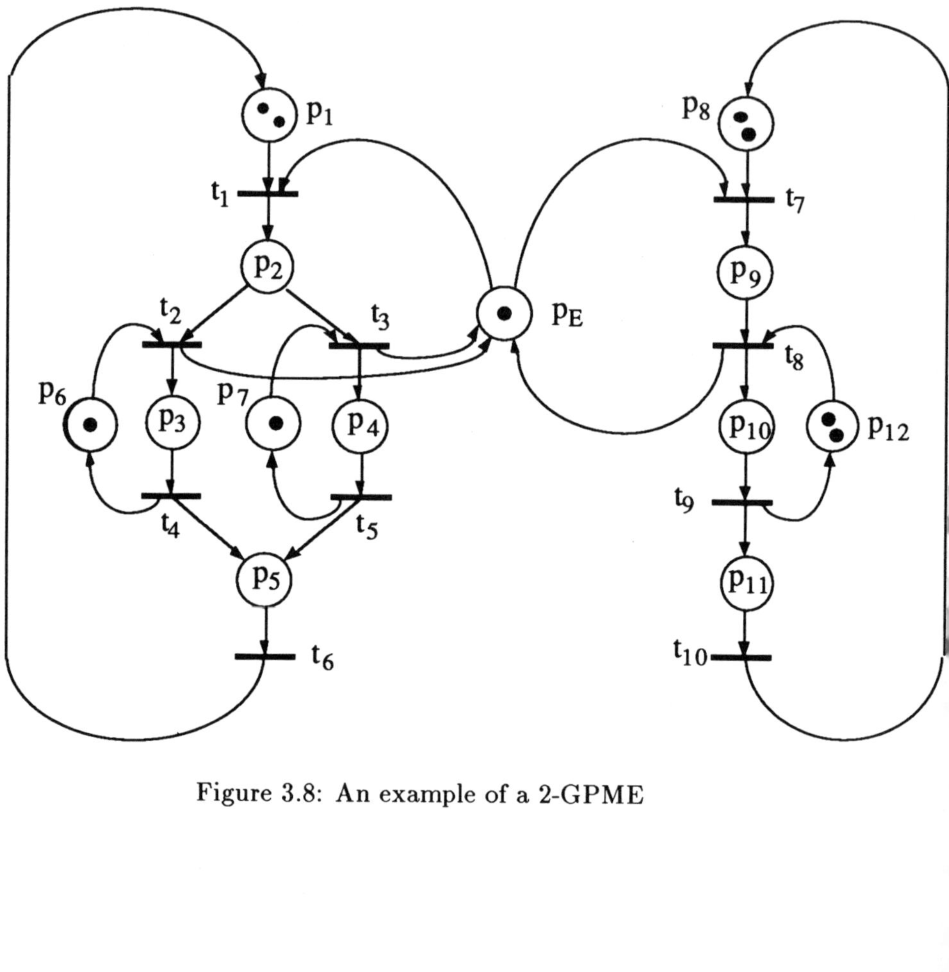

Figure 3.8: An example of a 2-GPME

Chapter 4

Sequential Mutual Exclusion

4.1 Introduction

Our study of Petri net synthesis for manufacturing systems begins with the classification of places in a Petri net model. The modeling approach in which places are used to model operation processes and the availability of resources, and in which transitions are used to model the start and/or end of operations makes such a distinction among places possible. Section 2.3 has divided places into three classes: operation places, fixed resource places, and variable resource places, called A, B, and C, respectively. Such a division depends upon a modeler's choice in many cases. A-places represent the operation processes and are not initially marked. B-places model the availability of fixed resources like robots and machines. C-places model the availability of variable resources such as fixtures or pallets, whose number needs to be specified. The number of the initial tokens in C-places can influence the qualitative properties of some systems. These systems contain resources which are shared by several sequential processes. If too many parts are allowed to enter such a system in a given period, the system may become deadlocked or blocked. Conceptualizing such phenomena, we define ***sequential mutual exclusion*** (SME) in the context of Petri net theory and find the underlying relationship between a net with such a structure and the net's initial marking over C-places. Also, to capture characteristics of a system with the resource shared by independent processes, we have defined and discussed the parallel mutual exclusion (PME) concept in Chapter 3. Each SME structure can be visualized as the sequential composition of several PMEs. The token capacity between two consecutive

PMEs is introduced and, roughly speaking, is the number of tokens which can be provided by the first PME without consumption by the second. Each token capacity is calculated in a reduced net which eliminates the shared resource. The results are used to determine the number of initial tokens in each C-place. This number can be interpreted as the number of jobs which can be dispatched to the system in the same interval without blocking the system.

This chapter will present the sequential mutual exclusion concept and the related theoretical results. The objectives of this chapter consist of:

1. Formulation of sequential mutual exclusion (SME) concept for a resource shared by sequential processes;

2. Formulation of token capacity and related concepts;

3. Derivation of the conditions under which Petri nets containing such structures are live, bounded, and reversible; and

4. To generalize these concepts and results for the Petri net modeling of more sophisticated automated manufacturing systems.

The next section starts with a manufacturing example and proceeds with the definition of SME. Then token capacity concept is formulated and discussed in Section 4.3 followed by discussion of the other useful parameters in Section 4.4. Theoretical results are developed in Section 4.5 followed by their further extension to more complicated cases in Section 4.6. Finally, SME is generalized to deal with choice structures and relevant results are presented in Section 4.7. We comment on the previous definitions and explanations of mutual exclusion and compare the work on shared resource allocation problems in computer science with our proposed PME and SME in Section 4.8.

4.2 Definition and Example

This definition deals with ***a sequential mutual exclusion*** in the sense that two transitions, e.g., t_{a1} and t_{a2}, can be enabled simultaneously only after t_{a1} fires from some initial marking before t_{a2}. A manufacturing example is given before a formal definition is presented.

Example 4.1: The manufacturing system is composed of three workstations, W_1, W_2, and W_3, and a robot (see Fig. 4.1). A part needs to be processed by W_1 first, then W_2, and finally W_3 to reach its final shape. Workstations W_1 and W_3 can each process only one part. W_2 can process two parts simultaneously and is able to load and unload itself. The robot, shared by W_1 and W_3, is used for loading and unloading these two workstations. Once

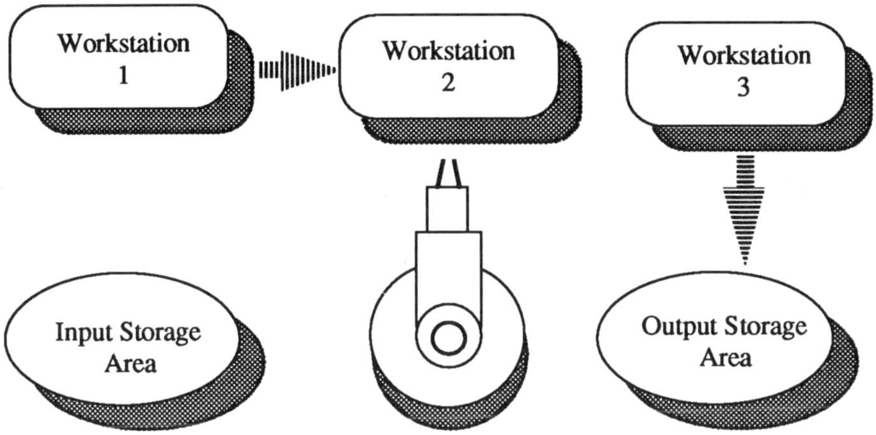

Figure 4.1: A three-workstation and one-robot manufacturing system

the robot starts loading either workstation, it cannot be interrupted until it finishes unloading the same one. Fixtured parts are awaiting processing in the input storage area. Final products will be automatically transported to the output storage area, whereas fixtures will be released to the input storage area by the robot as soon as the robot completes the unloading of W_3. There are four major steps which are modeled as four places in this system's Petri net model in Fig. 4.2.

Step 1 (place p_1) Fixtured parts in storage are available to W_1;

Step 2 (place p_2) The robot loads a part from the storage area; W_1 processes the part; the robot unloads W_1; the fixtured part is automatically moved to W_2.

Step 3 (place p_3) W_2 processes fixtured parts; a fixtured part is available to W_3.

Step 4 (place p_4) The robot loads a part to W_3; W_3 processes the part; the robot unloads W_3, and finally releases the fixture to storage.

The transitions represent either the start or completion of an operation process represented by each place. It should be noted that p_2, p_3, and p_4 may be further decomposed and expressed by a series of operation places. In this model, we have A-places $P_A = \{p_2, p_3, p_4\}$, which represent operation processes, B-places $P_B = \{p_5, p_6\}$ which represent resources with a fixed number, and a C-place $P_C = \{p_1\}$, which represents resources with a variable number.

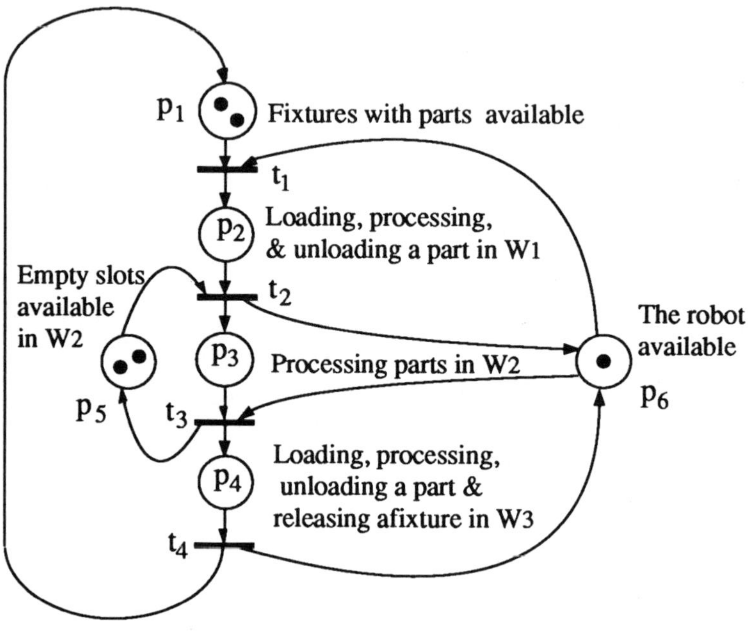

Figure 4.2: An example of sequential mutual exclusion

The robot is modeled as a shared resource place p_6 with $m_0(p_6) = 1$, and the availability of empty slots in W_2 is represented by p_5 with $m_0(p_5) = 2$. Suppose that p_1 is initially marked with two tokens to represent two unprocessed fixtured parts, i.e., $m_0(p_1) = 2$.

Initially, the robot can work for W_1 only. After a part goes through W_1 and W_2, the robot can work for either W_1 or W_3, but cannot work with both simultaneously. Using Petri net terminologies, t_1 should fire first, followed by t_2. Next t_3 may fire. However, even though they are both enabled at this marking $(1,0,1,0,1,1)^{\tau}$, t_1 and t_3 cannot fire simultaneously. This is a simple example of a sequential mutual exclusion whose definition follows.

Definition 4.1 Given a marked Petri net where $Z = (P,T,I,O,m_0)$ and $P = P_A \cup P_B \cup P_C$, *a sequential mutual exclusion* (SME) exists if and only if

1. $\exists (p_E, D)$ such that $D = D^1 \cup D^2 \cup \ldots \cup D^L$, $L \geq 2$, $k_i = |D^i|$, $D^i = \{(t^i_{aj}, t^i_{bj}), j \in N_{k_i}\}$, satisfying:
   (a) (p_E, D^i) is a k_i-PME in $Z_i = (P,T,I,O,m_0)$ that results from Z with $I(p_E,t) = O(p_E,t) = 0$, $\forall t \in D - D^i$;
   (b) $\forall t \in T^j_a, 1 \leq i < j, \exists t' \in T^i_a, \ni$ an A-path $EP(t',t) \neq \emptyset$ where $T^i_a = \{t^i_{av}, v \in N_{k_i}\}$;
   (c) if an A-path $EP(t^i_{bu}, t^j_{av}) \neq \emptyset$ for $u \in N_{ki}, v \in N_{kj}$, then $i < j$; and
   (d) if $EP(t^i_{bu}, t^j_{bv}) \neq \emptyset$ for $u \in N_{ki}, v \in N_{kj}$, $i < j$, then $t^j_{av} \in EP(t^i_{bu}, t^j_{bv})$;

2. $\forall m_0$, g_i, and $i \in N_{L-1}$, if $m_0[g_i>$ enables t^{i+1}_{aj}, $\forall j \in N_{k_{i+1}}$, then $\exists u \in N_{k_i}, \ni \#(g_i, t^i_{au}) = \#(g_i, t^i_{bu}) \geq 1$.

3. $\forall m_0$, if t^i_{av} fires at $m \in R(Z_i, m_0)$, then $\forall$ fireable g_j containing no t^i_{bv}, $\exists h_j$ containing no transition in T^j_a, $1 \leq j < i$ if $i > 1$, $\ni m[t^i_{av} g_j h_j >$ enables t^i_{bv} in Z_i.

In this definition, D^i, as well as D, is viewed as a set of transitions for convenience. D^i is also called the ith PME in the SME. Note that for two sets X and Y, $X - Y$ denotes the set $\{x : x \in X$ and $x \notin Y\}$ and other standard notation for handling sets is also used. Let,for $1 \leq i \leq L$,

$$T^i_a = \{t^i_{av}, v \in N_{k_i}\},$$

$$T^i_b = \{t^i_{bv}, v \in N_{k_i},$$

$$T_a = \cup_{1 \leq i \leq L} T^i_a,$$

and

$$T_b = \cup_{1\leq i\leq L} T_b^i.$$

Physically, an SME can model a resource shared by several sequentially related groups of processes. Each group contains independent processes and formulates a PME if other groups are ignored. The conditions in this definition are explained as follows.

Condition 1:

(a) (p_E, D^i) forms a k_i-PME if all the arcs of p_E related to transitions in $D - D^i$ are deleted.

(b) For each transition t in a PME with a higher index, there is a transition t' in a PME with a lower index such that an A-path exists between t' and t.

(c) There is no A-path from D^{i+1} to D^i; and

(d) Any elementary path containing t^i_{bu} and t^j_{bv} have t^j_{av} when $i < j$.

Our experience has suggested that all these requirements can be satisfied by many different manufacturing systems.

Condition 2: It guarantees that there exist sequential relations among different groups of processes.

Condition 3: It guarantees that each subnet Z_i has the property that once the transition t^i_{av} in D^i fires at a marking m, a sequence of transitions can be found to enable the other transition t^i_{bv} from any marking reachable from m in Z_i. It also implies that no preceding process, i.e., firing of transitions in $D^j, j < i$, is needed to finish the newly initiated process in D^i, whereas its post-process, i.e., firing of transitions in $D^v, v > i$ may need to be complete before the current one completes.

From the above definition, if $Z = (P_A \cup P_B \cup P_C, T, I, O, m_0)$ containing an SME (p_E, D), $D = D^1 \cup D^2 \cup \ldots \cup D^L$, then the following properties can be readily proved:

1. *If* $m_0[g_i >$ *enables* t^{i+1}_{aj}, $\forall j \in N_{k_{i+1}}$, *then for* $1 \leq v \leq i$, $\exists u \in N_{k_v}$, $\ni \#(g_i, t^v_{au}) = \#(g_i, t^v_{bu}) \geq 1$, $\forall i \in N_{L-1}$.

2. $\forall t \in D$, *if* $\exists p \in P_C$ *and an A-path* $EP(p,t)$, *then* $\exists t' \in T^1_a \cap EP(p,t)$, *then* $\exists t' \in T^1_a \cap EP(p,t)$.

The first property says that from an initial marking, a pair of transitions from a PME with a lower index in this SME have to fire before a transition in

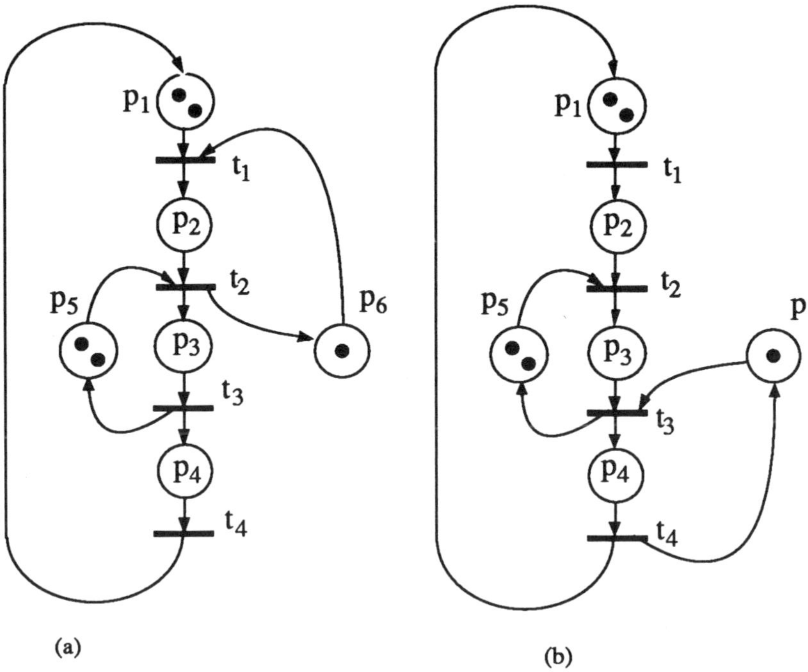

Figure 4.3: Z_1 and Z_2, resulting from Z with no consideration of the other PME

a PME with a higher index. This demonstrates the existence of a sequential relation among the PMEs.

The second property basically states that each transition in the concerned SME links a transition in its first PME by an A-path.

Example 4.2: In the Petri net model for the system in Fig. 4.2, a sequential mutual exclusion exists, i.e., $\{p_E, D^1 \cap D^2\}$ where $p_E = p_6$, $D^1 = \{(t_1, t_2)\}$ and $D^2 = \{(t_3, t_4)\}$. One can easily verify that it satisfies the proposed conditions in Definition 4.1 when $P_A = \{p_2, p_3, p_4\}$, $P_B = \{p_5, p_6\}$, and $P_C = \{p_1\}$. This is performed as follows.

1. Two subnets Z_1 and Z_2 are shown in Fig. 4.3. Since (p_E, D^1) and (p_E, D^2) are 1-PMEs in Z_1 and Z_2, respectively, and since in Z there is a unique A-path between t_1 and p_1, i.e., $t_1p_2t_2p_3t_3p_4t_4p_1$, Condition 1 is satisfied.

2. Since, in order to enable t_3, transitions t_1 and t_2 must fire first from any $m_0 = (m_0(p_1), 0, 1, 0, 1,)^\tau$ whenever $m_0(p_1) > 0$, Condition 2 is verified.

3. Consider Z_1 first. After t_1 fires, obviously transition t_2 can be enabled if there is tokens in p_5. Otherwise, p_3 has tokens and t_3 is enabled and can fire. Firing t_3 will mark p_5 and then t_2 is enabled. Consider $Z - 2$. After t_3 fires, transition t_4 is enabled immediately without firing any pre-process, i.e., t_1. Therefore, Condition 3 is verified.

The liveness and reversibility of the net can be affected if inappropriate initial tokens are distributed in a Petri net which contains a sequential mutual exclusion. For the example shown in Fig. 4.2, the system can be blocked when $m_0(p_1) \geq 3$. Specifically, when $m_0(p_1) = 3$, the dead state can be reached by firing $t_1 t_2 t_1 t_2 t_1$, and the dead marking is $(0, 1, 2, 0, 0, 0)^T$ for this net. In practical terms, the situation is that the workstations W_1 and W_2 are processing three parts, the robot is serving W_1, and W_2 is full already, while W_3 needs the robot to load a part from W_2.

The following observation can be made from the above example. In order to preclude the possibility of systems deadlocks, the total number of jobs which can enter a system in a particular time interval needs to be limited if there exist sequentially shared resources. One objective here is to determine number such that the system has no deadlock.

4.3 Token Capacity Concept

4.3.1 Definition

The system behavior of the Petri net discussed in Example 4.2 depends not only on the initial marking, but also on the processing capacity of Workstation W_2. This fact motivates us to propose a token capacity concept which can be used to study the influence of an SME upon the properties of the entire Petri net. This concept is related to the synchronic distance concept defined for some classes of Petri nets, but they are different in several respects, including motivation, applications, and results.

Definition 4.2 Given a marked Petri net $Z = (P, T, I, O, m_0)$, the *token capacity* $C(t, t')$ between transitions t and t' in Z is the maximum number of firings of t from the initial marking without firing t'.

Token capacity is identical to the concept of firing deviation bound which leads to the definition of synchronic distance [Silva 88]. From this definition, the value of $C(t,t')$ depends on both the initial marking and the net's structure. For the example shown in Fig. 4.3(a), if

$$m_0 = (m_0(p_1), 0, 0, 0, 2, 1)^T$$

then

$$C(t_1, t_3) = \min\{2, m_0(p_1)\}$$

$C(t,t) = 0$ is assumed. The token capacity may be difficult to evaluate for general cases. In these cases, the reachability graph may be used to calculate it. One property of the token capacity can be stated as follows:

Given t and t' in a marked Petri net Z, if there is an A-path $EP(t,t')$ and if for any transition $r \in T$, the first intersection of $EP(r,t')$ and $EP(t,t')$ is a transition, then $C(t',t) = 0$.

This property states that from an initial marking, if firing t' requires firing t first, then $C(t',t) = 0$. In other words there is a strict sequential relationship between t and t'. This capacity can be used to determine whether there is a sequential relationship between any two transitions given a marked Petri net.

4.3.2 Token Capacity of an SME

In this section the token capacity between two consecutive PMEs in an SME is defined. However, we must first introduce the following notation.

Suppose that $Z' = (P', T', I', O', m_0')$, and $Z = (P, T, I, O, m_0)$. Z is called *a subnet* of Z' with P and T if $P \subset P'$, $T \subset T'$, and if, $\forall p \in P$, $t \in T$, $I(p,t) = I'(p,t)$, $O(p,t) = O'(p,t)$, $m_0(p) = m_0'(p)$. We also write $Z \subset Z'$.

Assume that $Z = (P_A \cup P_B \cup P_C, T, I, O, m_0)$ is a subnet of $Z' = (P', T', I', O', m_0')$. $P' = P_A \cup (P_B \cup \{p_E\}) \cup P_C$, $T' = T$, and $D = D^1 \cup D^2 \cup \ldots \cup D^L$, where $k_i = |D^i| \geq 1$, $L \geq 2$, $D^i = \{(t^i_{a1}, t^i_{b1}), (t^i_{a2}, t^i_{b2}), \ldots, (t^i_{ak_i}, t^i_{bk_i})\}$, $i \in N_L$. (p_E, D) is an SME in Z'. Z is the net which has the same transitions that Z' has, but contains neither p_E nor its arcs related to transitions in Z'. Also, assume that Z' satisfies Assumptions 2.1 to 2.2.

For two consecutive PMEs D^i and D^{i+1} in Z', consider a pair of transitions t and t', $t \in T^i_b$ and $t' \in T^{i+1}_a$. If an elementary path $EP(t,t')$ exists, i.e., $EP(t,t') \neq \emptyset$, then $Z^e(t,t') = (P^e, T^e, I^e, O^e, m^e_0)$ is, roughly speaking, a subnet of Z with two kinds of the places and transitions. One is the set of nodes that belong to an elementary circuit that starts, ends at a B-place, and contains an elementary path between t and t'. Another is the set of nodes that do not belong to any elementary circuits including a C-place, but are related to t or t'. In other word, we remove all C-places, the shared resource place p_E, and the places and transitions unrelated to any elementary circuits which include t and t' in $Z^e(t,t')$. It is obvious that $Z^e(t,t') \subset Z \subset Z'$ and m_0^e is unique because all C-places are removed in $Z^e(t,t')$.

Given $p \in P_C$, $Z^a(p) = (P^\alpha, T^\alpha, I^\alpha, O^\alpha, m_0^\alpha)$ is defined as a subnet of Z' with $P^\alpha = P' - \{p_E\} - (P_C - \{p\})$, and $T^\alpha = T'$ and with $O^\alpha(p,t) = 0$, $t \in T^\alpha$. Strictly, $Z^\alpha(p)$ is not a subnet of Z' from the definition. Informally, $Z^\alpha(p)$ is the net Z' with the removal of p_E and other C-places and the arcs to p from any transitions. It will be used later to define three parameters α, β and γ.

Summarizing, the following definition gives a complete description of the above three kinds of subnets.

Definition 4.3 Given a Petri net Z' that contains an SME structure, three kinds of subnets are defined as follows:

1. $Z = (P, T, I, O, m_0)$ is a subnet of Z' with $P = P' - \{p_E\}$, $T = T'$
2. Given t, $t' \in T$, $t \neq t'$, $Z^e(t,t') = (P^e, T^e, I^e, O^e, m_0^e)$ is a subnet of Z with $P^e = (P \cap EC_{t,t'}) \cup (P - \cup_{p_c \in P_C} EC(p_c) - \{p \in P : EP(t,p) = \emptyset$ & $EP(p,t) = \emptyset\})$
   and
   $T^e = (T \cap EC_{t,t'}) \cup (T - \cup_{p_c \in P_C} EC(p_c) - \{r \in T : EP(t,r) = \emptyset$ & $EP(r,t) = \emptyset\})$,
   where
   $EC_{t,t'} = \{EC(p) : p \in P_B$, and $\exists x \in EP(t,t')$, $\ni x \in EC(p)\}$.
3. Given $p \in P_C$, $Z^\alpha(p) = (P^\alpha, T^\alpha, I^\alpha, O^\alpha, m_0^\alpha)$ is a subnet of Z' with $P^\alpha = P' - \{p_E\} - (P_C - \{p\})$, and $T^\alpha = T'$ and with $O^\alpha(p,t) = 0$, $t \in T^\alpha$.

Note that $EC_{t,t'}$ is the set of all nodes which are in the circuits containing a B-place and at least a node from one elementary path $EP(t,t')$.

Example 4.3: Suppose that Z' is the net discussed in Fig. 4.2. $Z^e(t_2,t_3)$, Z, and $Z^\alpha(p_1)$ are constructed as shown in Fig. 4.4(a), (b), and (c), respectively. It should be noticed that $Z = Z^\alpha$ when there exists one C-place only.

Definition 4.4 Assume that Z', Z, and $Z^e(t,t')$ are specified as Definition 4.3. Given t and t', $t \in T_b^i$ and $t' \in T_a^{i+1}$,

1. If $EP(t,t') \neq \emptyset$, $C(t,t')$ in $Z^e(t,t')$ is *achievable with respect* to $Z^e(t,t')$ if
   (a) $\exists g$, $\ni \#(g,t) = C(t,t')$, $\#(g,t') = 0$ in $Z^e(t,t')$; and

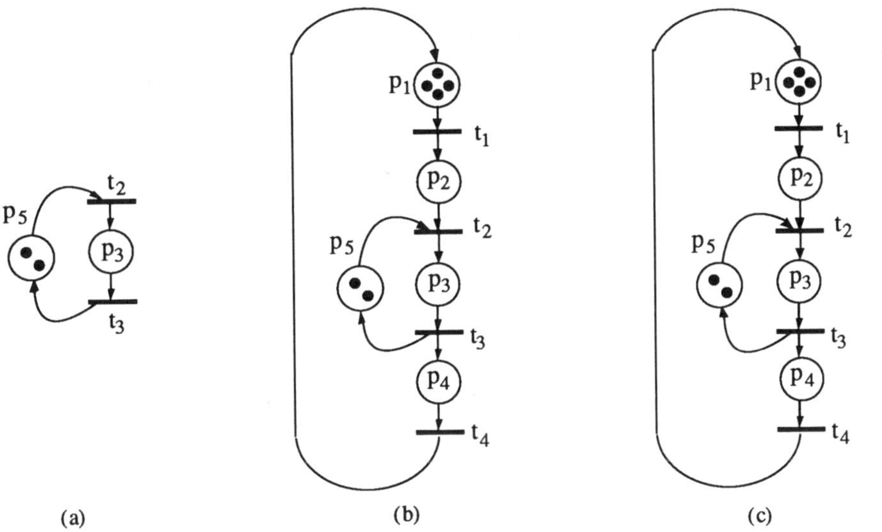

Figure 4.4: Construction of (a) $Z^e(t_2, t_3)$, (b) Z, and (c) $Z^\alpha(p_1)$ from Z' shown as in Fig. 4.2

(b) $\exists t'' \in T_a^i \cap T^e$, and h, $\ni$ an A-path $EP(t'',t) \neq \emptyset$, $m_0[gh>$ enables t'' in $Z^e(t,t')$.

Otherwise, $C(t,t')$ is ***unachievable with respect to*** $Z^e(t,t')$.

2. If $EP(t,t') \neq \emptyset$, $C(t,t')$ in $Z^e(t,t')$ is ***achievable with respect to*** Z if

   (a) $\exists m_0$ and g, $\ni \#(g,t) = C(t,t')$, $\#(g,t') = 0$ in Z; and

   (b) $\exists t'' \in T_a^i$, and h, $\ni$ an A-path $EP(t'',t) \neq \emptyset$, $m_0[gh>$ enables t'' in Z.

   Otherwise, $C(t,t')$ is unachievable.

3. $C(t,t')$ is *effective* if $Z^e(t,t')$ is reversible and $\forall p_b \in P^e \cap P_B$ and $p_c \in P_C$, if $\exists r \in EP(p_c,p_b) \cap EPD$, then either $t \in EP(p_c,p_b)$ or $t' \in EP(p_c,p_b)$, where EPD in Z is the set of all elementary paths between any two transitions in D; Otherwise it is *ineffective*.

If $C(t,t')$ in $Z^e(t,t')$ is achievable with respect to $Z^e(t,t')$, then the system may be deadlocked no matter how we distribute the initial tokens to the other C-places. For example, suppose that the C-place p_1 in the Petri net as shown in Fig. 4.2 becomes a fixed resource place or B-place marked initially with four tokens. Then, the net $Z^e(t_2,t_3)$ equals to the subnet Z shown in Fig. 4.4 (b). Token capacity $C(t_2,t_3)$ is achievable with respect to $Z^e(t_2,t_3)$. Obviously, the net in Fig. 4.2 can be deadlocked when $m_0(p_1) = 4$.

Practically, when such a case is encountered, we can always find a B-place which can be viewed as a C-place in our design process. Therefore, we can assume that $C(t,t')$ is unachievable with respect to $Z^e(t,t')$ in order to have a meaningful discussion. The assumption follows.

Assumption 4.1 Given t and t' in Z', $t \in T_b^i$ and $t' \in T_a^{i+1}$, $C(t,t')$ in $Z^e(t,t')$ is unachievable with respect to $Z^e(t,t')$.

In the following discussion, we say that a token capacity is achievable or unachievable when it is achievable or unachievable with respect to Z.

Assume that a set of B-places contribute their tokens to the token capacity $C(t,t')$ in $Z^e(t,t')$. Then the achievability of the token capacity with respect to Z assumes that the resources can be fully used, while the unachievability often implies a poor design because of possibly insufficient use of the resources, thereby degrading the system performance. However, a designer may use this to design a system such that the number of the jobs which can enter a critical part of the system can be limited and the deadlock-freeness is guaranteed. For example, design the system such that $m_0(p_1)$ is fixed to one, i.e., only one part is allowed to be present in the system each time. The system is certainly deadlock-free. However, the system performance is degraded because no pipeline concurrency is possible. For most manufacturing system applications, the token capacities are achievable.

The effectiveness of $C(t,t')$ guarantees that the tokens in those B-places can become available from any marking without firing any transitions in the SME in Z'. The common case is that all these B-places represent non-shared resources. Then the above defined net $Z^e(t,t')$ is obviously reversible and any path between a C-place and the B-place has to contain either t or t' in Z. Therefore, $C(t,t')$ is effective. For example, $Z^e(t_2,t_3)$ is such a case shown in Fig. 4.4(a) and is reversible.

The effectiveness of $C(t,t')$ precludes the cases where a fixed resource place p_b models many identical resources shared by a number of parallel sequentially related processes. These processes are initiated due to the tokens originating from different C-places. Hence, the empty slots guaranteed by the the resource place p_b are consumed by tokens in different C-places. For example, a shared buffer in a manufacturing may constitute such a case. The method of dealing with a Petri net containing such shared buffers is discussed in Chapter 5 as an extension of the current theory.

In the following discussion, $C(t,t')$ is assumed to be effective. When a practical Petri net is given, it can be verified to be effective unless otherwise stated.

Definition 4.5 Assume that $C(t,t')$ is effective if $\exists j \in N_{L-1}$, $t \in T_b^j$, and $t' \in T_a^{j+1}$, $\ni$ A-path $EP(t,t') \neq \emptyset$. The token capacity between D^i and D^{i+1} in Z' is defined as

$$C^i = \min_{1 \le j \le k_i} \{C_j^i, i \in N_{L-1}\},$$

where

$$C_j^i = \min\{CC(t_{bj}^i, t'), t' \in T_a^{i+1}\}$$

$$CC(t_{bj}^i, t') = \begin{cases} \infty & \text{if } EP(t_{bj}^i, t') = \emptyset \text{ in } Z \\ C(t_{bj}^i, t') & \text{if } C(t_{bj}^i, t') \text{ in } Z^e(t_{bj}^i, t') \text{ is achievable} \\ \infty & \text{otherwise} \end{cases}$$

The token capacity between two consecutive PMEs can be understood as the capacity to hold the jobs or parts between two group of processes. Such a capacity often depends on a buffer capacity or machine processing capacity. In practice, such capacities are *effective* for most manufacturing systems. When $L = 2$, we also call C^1 the token capacity of the SME.

Example 4.4: $\{p_6, D^1 \cup D^2\}$, where $D^1 = \{(t_1, t_2)\}$ and $D^2 = \{(t_3, t_4)\}$, is an SME in the Petri net shown in Fig. 4.2. $Z^e(t_2, t_3)$ and Z can be obtained as depicted in Fig. 4.4. From Fig. 4.4(a), $C(t_2, t_3) = 2$. This is achievable in Z because t_1 can fire numerous times when $m_0(p_1)$ is large enough even if t_4 does not fire at all in Z. This is also effective because $Z^e(t_2, t_3)$ is reversible and either path between p_1 (unique C-place) and p_5 (unique B-place in $Z^e(t_2, t_3)$) contains t_3. Therefore, $C^1 = 2$. Here C^1 is the part processing capacity of Workstation W_2.

It should be indicated that the computational complexity of $C^i = C(t, t')$ is significantly alleviated because of the removal from the original net of (a) the shared resource place p_E and its arcs to the transitions, and (b) places, transitions, and arcs between P_C and any elementary circuit including t and t'. The token capacities between two consecutive PMEs are later used to determine the maximum number of tokens which can be initially distributed over its C-places so that the desired properties of the net can be preserved.

4.4 Parameters α, β, and γ

The token capacity of an SME is not enough to describe the system properties of a Petri net containing SME structures. The following parameters are introduced based on the Petri net $Z^\alpha(p)$ defined in the last section where p is a C-place. It is in fact a net Z' with the removal of p_E and other C-places and the arcs to p from any transitions.

Definition 4.6 Given Z' and $Z^\alpha(p)$, $p \in P_C$ as defined in Definition 4.3, for $t \in T_a^i$, $1 \le i \le L-1$,

$$\alpha^i(p,t) = \min\{m_0(p) : t \text{ can be enabled in } Z^\alpha(p)\}$$

$$\beta^i(p,t) = \begin{cases} 0 & \text{if } \alpha^i(p,t) = 0; \\ \min\{x : x m_0(p) \ge \gamma^i(p,t,m_0(p))\} & \text{Otherwise.} \end{cases}$$

where $\gamma^i(p,t,m_0(p)) = \max$ {number of firings of t in $Z^\alpha(p)$ given $m_0(p)$}.

Note when no confusion arises:

$$\alpha^i(p) = \max\{\alpha^i(p,t), t \in T_a^1\}$$

$$\beta^i(p) = \max\{\beta^i(p,t), t \in T_a^1\}$$

$$\gamma^i(p,t) = \max\{\gamma^i(p,t,m_0(p)), m_0(p) > 0\}$$

and

$$\gamma^i(p) = \min\{\gamma^i(p,t), t \in T_a^1\}.$$

Since we focus on the system with an SME of $L = 2$, we discard the superscript i of these parameters in the following discussion although the given comments can be applied to an SME with $L > 2$. When we deal with the case $L = 2$, or the case $L > 2$, $\alpha^i = \alpha$, $\beta^i = \beta$, and $\gamma^i = \gamma$ for $1 \leq i \leq L-1$, only α, β and γ will be used.

Explanation of α

For $p \in P_c$ and $t \in T_a$, $\alpha(p,t) = 0$ implies that enabling t has nothing to do with p if there are enough tokens in other C-places. This is true when an operation process requires none of the resources represented in the place p.

$\alpha(p,t) \neq 0$ implies that $\alpha(p,t)$ tokens in p are needed to fire t. Practically, the operation initiated by firing transition t needs at least $\alpha(p,t)$ resources available in the variable resource place p. $\alpha(p)$ represents the right number of tokens required in p in order to possibly fire any transition in the SME. Otherwise if the number of tokens in p is less than this number, some transition in the SME is never enabled. It is easy to see that $\alpha(p)$, $\forall p \in P_C$, represents the minimum resource requirement to operate a real-time manufacturing system.

Explanation of β

For $p \in P_c$ and $t \in T_a$, $\beta(p,t) = 0$ has the same explanation as $\alpha(p,t) = 0$. It offers a measure of how fast the number of firings of a transition t increases as the initial tokens increases in p. The larger $\beta(p,t)$, the more the firings of t. Generally, the relationship between $m_0(p)$ and $t's$ firings is linear, or piecewise linear. Therefore, $\beta(p,t)$ exists.

$\beta(p,t) = 1$ for most cases, but for some important manufacturing assembly cases when choice-synchronization structures are used (see Section 5.1.2 for an example), $\beta(p,t)$ may be less than 1. Because of Assumption 2.2 which states that two tokens from different C-places are not allowed to flow into an identical place without any synchronization operation, $\beta(p,t)$ always exists if the net is bounded. For a live Petri net, it can be proved that $\beta(p,t) > 0$.

$\beta(p) = 0$, where p is a C-place, means that $m_0(p)$ contributes nothing to the firing of any transitions in the concerned SME in the Petri net. Therefore, $m_0(p)$ could be any positive number. $0 < \beta(p) < 1$ assumes that initial tokens in p need be divided so as to fire several transitions that may be in a conflict.

$\beta(p) = \infty$ means that Z with no SME resource place p_E will produce some tokens which play the same role as initial tokens in p. Such a system is impractical and only of theoretical interests since Z will be unbounded if it is live.

Explanation of γ

For $p \in P_c$ and $t \in T_a$, $\gamma(p,t,m_0(p))$ indicates the maximum number of firings of t given $Z^\alpha(p)$ and $m_0(p)$. Hence, $\gamma(p,t,m_0(p)) = 0$ means that t cannot be enabled. $\gamma(p,t,m_0(p)) = \infty$ means $\beta(p) = \infty$ and therefore, implies that the system is not practical. Often, $\gamma(p,t,m_0(p))$ is a linear function of $m_0(p)$.

Finite $\gamma(p,t)$ means that t's firings are limited to a certain number no matter how many initial tokens are in p. $\gamma(p,t) = 0$ implies that neither Z' nor Z is live. However, $\gamma(p,t)$ is often infinite and this implies that the operations can be performed provided that there are infinite resources available if no limitation can be imposed by other fixed resources represented by B-places.

$\gamma(p)$ is finite implies that there is another fixed number of resources which confine the execution of some operation processes. The common case is that $\gamma(p) = \infty$.

These parameters become important later in the discussion of the influence of an SME upon the liveness and reversibility of a Petri net.

Example 4.5: From the Petri net as shown in Fig. 4.2, we construct Petri net $Z^\alpha(p_1)$ as depicted in Fig. 4.4(c). Then,

$$\alpha(p_1) = \alpha(p_1,t_1) = 1$$

$$\beta(p_1) = \beta(p_1,t_1) = 1$$

and

$$\gamma(p_1,t_1,m_0(p)) = m_0(p).$$

Therefore,

$$\gamma(p_1,t_1) = \gamma(p_1) = \infty.$$

4.5 Development of Theoretical Results

The following assumptions are made for Lemmas 4.1 to 4.5, and Theorems 4.1 to 4.4 which deal with a Petri net having sequential mutual exclusions.

1. Given $Z' = (P',T',I',O',m_0')$ with $P' = P_A \cup (P_B \cup \{p_E\}) \cup P_C$ and $T' = T$, $Z = (P_A \cup P_B \cup P_C, T, I, O, m_0)$ is a subnet of Z'.

2. (p_E, D) is an SME in Z' where $D = D^1 \cup D^2 \cup \ldots \cup D^L$ where $k_i = |D^i| \geq 1$, $L \geq 2$, $D^i = \{(t^i_{a1}, t^i_{b1}), (t^i_{a2}, t^i_{b2}), \ldots, (t^i_{ak_i}, t^i_{bk_i})\}$, $i \in N_L$.

It can be shown that there exist a series of dead transitions in a Petri net with an SME when the number of initial tokens of each C-place exceeds some limit if C^i is finite and achievable, $\forall i \in N_{L-1}$. Theorem 4.1 states this by using β^i that describes the relations between C-places and the SME. Before it is given, we prove a lemma first.

Lemma 4.1 *If the token capacity of an SME of Z', C^i exists, then $\gamma^i(p) \geq C^i$ where $\gamma^i(p)$ is defined in Definition 4.6.*

Proof: From the definitions of C^i and $\gamma^i(p)$, the lemma follows.

Theorem 4.1 *If C^i is the finite token capacity of the SME in Z' and is achievable, then there is a finite $M > 0$ such that if $m_0(p) > M$, $\forall p \in P_C$, every transition in the SME is not live.*

Proof: This indicates that the token capacity between D^i and D^{i+1} is C^i. Without loss of generality, suppose that at most C^i tokens can exist on EP which represents all A-paths between, for example, t^i_{b1} and t^{i+1}_{a1}. Since C^i is achievable, t^i_{b1} can fire C^i times without firing t^{i+1}_{a1} so that there are C^i tokens staying on EP. Let $M > [C^i/\max\{\beta^i(p) : p \in P_C\}]$. Since $m_0(p) > M$, $\forall p \in P_C$, $\beta^i(p)m_0(p) > C^i$ if $\beta^i(p) > 0$. Therefore, we can fire t^i_{a1} and other related transitions in $EP(T^i_a, T^i_b)$ since the token capacity C^i is achievable. After that, however, t^i_{b1} cannot fire because C^i is the maximum number of tokens on EP. None of t^{i+1}_{a1} and t^i_{a1} can fire either because there is no token in p_E. All transitions related to p_E, i.e, ones in the SME, are dead from this marking. Therefore, there is a set of dead transitions in this Petri net Z. QED

Theorem 4.1 shows the fact that a system can be at least partially deadlocked, i.e., the shared resource and its related machines cannot process any further jobs when improper initial tokens are distributed over C-places. When the deadlocked portion is critical for the whole system, for example, when each job needs to go through that portion, then the system is completely deadlocked.

However, this property does not claim whether Z is deadlock-free or not if $\beta^i(p)m_0(p) \leq C^i$. The next theorems deal with the conditions that make a Petri net containing an SME bounded, live, and reversible.

Lemma 4.2 *p_E is a safe place in Z'.*

The proof is similar to that of Lemma 3.1 in the last chapter.

Theorem 4.2 *Z' is bounded (safe) if Z is bounded (safe).*

Proof: The proof is similar to that of Theorem 3.1. QED

Before Lemmas 4.3-4.5 are given, we formulate Condition C0 called unachievability condition of the token capacity of the SME. The condition states that whenever there is an A-path between two transitions t and t', $t \in T_b^i$, $t' \in T_b^{i+1}$, $1 \leq i \leq L-1$, then finite token capacity $C(t,t')$ is effective and unachievable.

C0. Token capacity $C(t_{bu}^i, t_{av}^{i+1})$ is effective and unachievable if an A-path $EP(t_{bu}^i, t_{av}^{i+1})$ in Z exists for $1 \leq i \leq L-1$, $1 \leq u \leq k_i$, and $1 \leq v \leq k_{i+1}$, and if $C(t_{bu}^i, t_{av}^{i+1})$ is finite.

Lemma 4.3 *If C0 is true and Z is live, then after t fires at m, $\forall$ fireable g, $\exists h, \ni m[tgh>$ enables t' in Z', where $(t,t') \in D^i$, $1 \leq i \leq L$.*

Proof: Suppose $m \in R(Z', m_0')$, and $t \in T_a^i$ is enabled at m. Firing t and g, we get: $m[tg> m'$. Then, $m(p_E) = 1$ and $m'(p_E) = 0$. Since firing t removes the token from p_E, g contains no transition in T_a in D. At m', any transition in T_a cannot be enabled.

Case 1: $i = L$. According to the definition of the SME, since Z is live, Z_L is live using Theorem 3.2 where Z_L is the Z' which excludes arcs between P_E and $D - D^L$. In Z_L, t' can be enabled without firing any transitions in $D - D^L$, such a sequence exists for Z'. Therefore, t' can be enabled in Z'.

Case 2: $i < L$. There exists D^{i+1}. If there is no $t'' \in T_a^{i+1}, \ni EP(t',t'') \neq \emptyset$, then the case is equivalent to $i = L$. Otherwise, $\exists t'' \in T_a^{i+1}, \ni EP(t',t'') \neq \emptyset$. The two possibilities are: $C(t',t'') = \infty$ and $C(t',t'') < \infty$. When $C(t',t'') = \infty$, this assumes that t' can be enabled after t fires and any other transitions fire in Z_i which is live. Such a sequence also makes t' enabled in Z'. When $C(t',t'') < \infty$, since $C(t',t'')$ is effective and unachievable, t' can also be enabled after t fires and any other transitions fire in Z_i which is live. Such a sequence also makes t' enabled in Z'. This accomplishes the proof. QED

Lemma 4.4 *If C0 is true, then Z' is live with respect to $m_0' = (m_0, 1)$ if Z is live with respect to m_0.*

A complete proof of this lemma is given in Appendix A. We explain its significance as follows. The condition C0 says that any finite token capacity between the two transitions from two consecutive PMEs in an SME is effective, i.e., resources can become available whenever they are needed. The unachievability of the token capacity in the condition implies that enough resources are available to start any operations at any marking. Intuitively, such a system should be live due to the redundant resources available if the original system with no shared resource is live. Using induction, we prove this lemma by constructing a desirable sequence of transitions based on a given one which is fireable in the known live net. Using the method similar to that of Lemma 4.4, we can show the following lemma.

Lemma 4.5 *If C0 is true, then Z' is reversible with respect to $m_0' = (m_0, 1)$ if Z is reversible with respect to m_0.*

The following conditions are needed for the two following theorems where $\alpha(p) = \max\{\alpha^i(p) : 1 \leq i \leq L-1\}$:

C1. $\exists C^i$, m_0, α, and β^i , $\ni m_0(p) \geq \alpha(p)$, and $\beta^i(p)m_0(p) \leq C^i, \forall p \in P_C$;

C2. $\forall m \in R(Z, m_0)$, $m(p) \leq m_0(p)$ if $p \in P_B \cup P_C$, and

C3. if an A-path $EP(t^i_{bu}, t^{i+1}_{bv})$ in Z exists for $1 \leq i \leq L-1$, $1 \leq u \leq k_i$, and $1 \leq u \leq k_{i+1}$, and if $C(t^i_{bu}, t^{i+1}_{bv})$ is finite, then $C(t^i_{bu}, t^{i+1}_{bv})$ is effective.

Before Theorems 4.3 and 4.4 are given, these conditions are explained as follows.

1. C1 defines a range of the initial marking. Practically, it conveys information, for example, on how many jobs are allowed to be dispatched to the system in any given time interval.

2. C2 states that the Petri net without this SME never produces more than the initially given tokens in those resource places at any reachable markings.

3. C3 requires that the token capacity between any two consecutive PMEs be *effective* in use if it exists as a finite number.

For a practical manufacturing system, these conditions should be true and, therefore, be satisfied by its Petri net model.

Theorem 4.3 *Z' is live if Z is live and C1, C2, and C3 are true.*

Proof: In order to use Lemma 4.4 to prove this theorem, construct two new Petri nets as follows. Given any marking satisfying C1, let

$$\hat{Z} = (P_A \cup \hat{P_B} \cup \hat{P_C}, T, I, O, m_0),$$

and

$$\hat{Z'} = (P_A \cup (\hat{P_B} \cup \{P_E\}) \cup \hat{P_C}, T, I, O, m_0),$$

where $\hat{P_B} = P_B \cup P_C$, $\hat{P_C} = \emptyset$, and all the others are as same as those of Z and Z'. Conditions C1, C2, and C3 make token capacities involved in the SME in $\hat{Z}'$ effective and unachievable. Also, $\hat{Z}$ is live since Z is live. According to Lemma 4.4, $\hat{Z}'$ is live. Therefore, Z' is live. Theorem 4.3 follows. QED

Similarly, using Lemma 4.5 we can prove the following theorem.

Theorem 4.4 *Z' is reversible if Z is reversible and C1, C2, and C3 are true.*

Theorems 4.2-4.4 give only sufficient conditions and one can also easily provide examples such that the necessity of all theorems is not true as we did for Theorems 3.1-3.3 regarding parallel mutual exclusions.

Example 4.6: Consider the Petri net shown in Fig. 4.2, where $P_C = \{p_1\}$. As we computed before, $C^1 = 2$, $\beta(p) = 1$, and $\alpha(p) = 1$. Since Conditions C1 to C3 are obviously satisfied, the net in Fig. 4.2 is bounded, live, and reversible when $1 \le m_0(p_1) \le C^1 = 2$ by Theorems 4.3-4.5.

Suppose that only one empty slot is available in Workstation W_2; i.e., $m_0(p_5) = 1$. Now, $C^1 = 1$. Then the net is safe, live, and reversible only when $m_0(p_1) = 1$ using Theorems 4.2 to 4.4. This is true because the only possible firing sequences are $(t_1t_2t_3t_4)^n t_1t_2t_3$, $(t_1t_2t_3t_4)^n t_1t_2$, $(t_1t_2t_3t_4)^n t_1$, and $(t_1t_2t_3t_4)^n$ where $n \in N$.

Example 4.7: Suppose that the two nets Z_a and Z_b are given as shown in Fig. 4.5. Both nets have the same transitions and places:

$$P_A = \{p_2, p_3, p_4, p_5\}$$

$$P_B = \{p_6, p_7\}$$

$$P_C = \{p_1\}.$$

They also have the same structure except that Z_a has the arc (p_6, t_2) whereas Z_b has the arc (p_6, t_1). One can easily verify that both nets contain an SME, i.e., $(p_7, \{(t_1, t_3\} \cup \{t_4, t_5\})$.

Two subnets to compute token capacities of the SMEs are shown in Fig. 4.6. For both nets we have

$$C(t_3, t_4) = 1.$$

Using Theorems 4.3 and 4.4, we have the results:

if $m_0(p_1) = 1$, both nets are live and reversible.

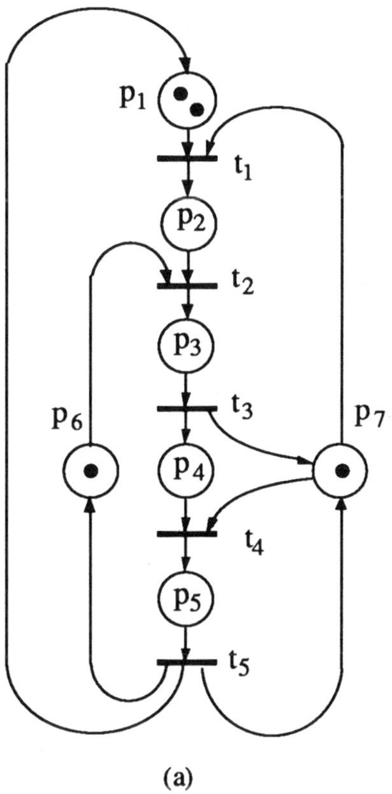

Figure 4.5: Two Petri nets with different qualitative properties: (a) Z_a is neither live nor reversible, and (b) Z_b is live and reversible

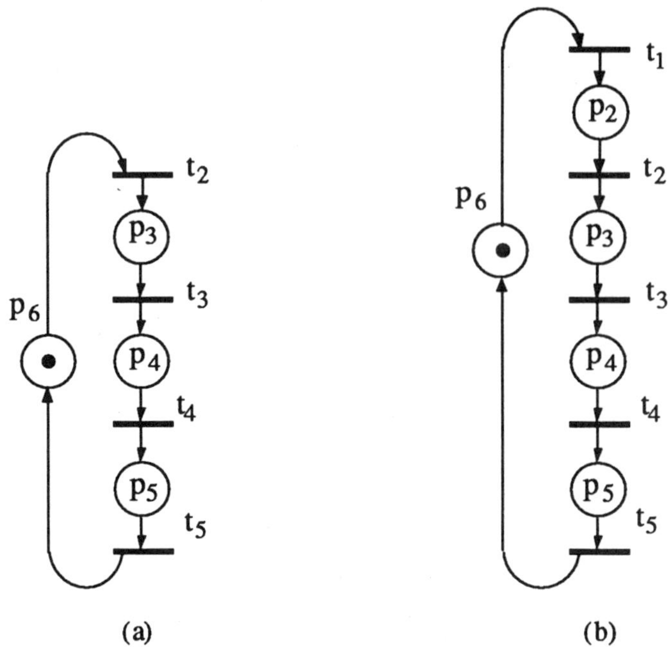

Figure 4.6: Two subnets (a) $Z^e(t_2, t_3)$ for Z_a, and (b) $Z^e(t_2, t_3)$ for Z_b to compute token capacities of the SMEs

After t_3 fires in Z which results from Z_b with the removal of the shared resource place, p_7, t_1 in the SME cannot be enabled without firing t_4 and t_5. Therefore, $C(t_3, t_4)$ is unachievable in Z_b. Using Lemmas 4.4 and 4.5 we conclude that

whenever $m_0(p_1) > 0$, the net Z_b is live and reversible.

However, when $m_0(p_1) = 2$, Z_a can be deadlocked. The deadlock occurs after we fire $t_1t_2t_3t_1$. We can easily verify such results for both nets using the reachability graph. The example shows that minor difference in the nets' structures can lead to their different behaviors. Such phenomena will be further discussed in the following section.

4.6 Further Theoretical Extensions

There are two cases which need to be further clarified for sequential mutual exclusions. One occurs when token capacities among different consecutive PMEs become unachievable. The other occurs when the token capacities result from the B-places that are shared by the related PMEs in the same SME. In other words, the token capacities are no longer effective. Discussions and results for both cases are presented.

4.6.1 Unachievable Token Capacities

Studying the two Petri nets presented in Fig. 4.5, we find that the system properties of the first net Z_a is very sensitive to a minor change in initial token distribution over its C-place. For example, when we increase $m_0(p_1)$ from 1 to 2, Z_a is neither live nor reversible. By comparison, the net Z_b has the desirable properties which are unchanged no matter how many tokens are initially in p_1. Such difference results from different properties of token capacities of the two SMEs. Since the token capacity of the SME in Z_b is unachievable, using Lemmas 4.4 and 4.5 guarantees that the net Z_b is live and reversible whenever $m_0(p_1) \geq 0$.

Hence, designers may adopt a strategy which guarantees that every token capacity related to an SME is unachievable. The resulting net has desirable properties which are not sensitive to any increase of initial tokens in a C-place. Such a strategy can, in fact, be developed as an approach to modeling a deadlock-free control structure for a manufacturing system where a very conservative resource allocation policy is used. This policy can be summarized thus:

If any process involves several resources, it can start only after all these resources are available.

Based on the above principle, one may design a system without any deadlocks under various initial token distributions. The theoretical foundation for the results from such a policy is in part accomplished in this book.

Such a policy can be inefficient in distributed systems [Singhal 89], where numerous processes and resources exist and deadlocks may be neither frequent and nor costly to detect and resolve. However, in the manufacturing environment deadlocks are often expensive to detect and resolve. Therefore, deadlock-free control structures may be desirable. The trade-off issues between system deadlock and performance need to be studied [Zhou 90d].

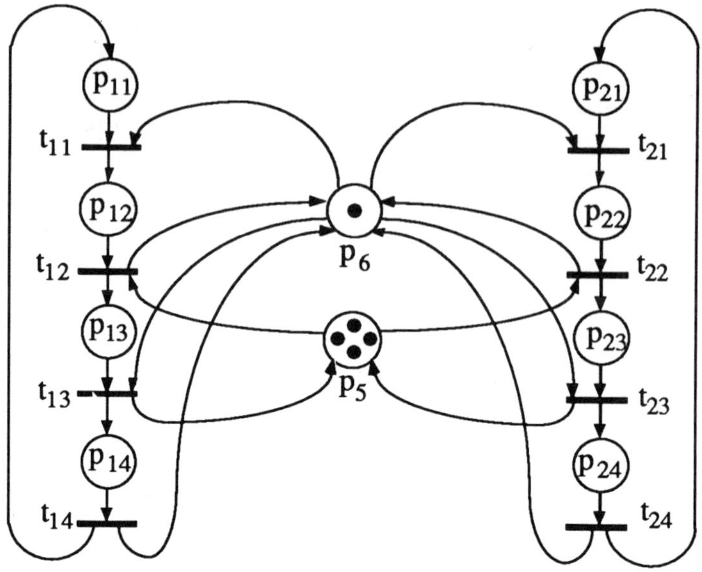

Figure 4.7: A Petri net with ineffective token capacity between the two PMEs in the SME

4.6.2 Ineffective Token Capacities

To show the problem at hand, we consider the Petri net as shown in Fig. 4.7. Without much consideration of physical implementation, the robot (p_6) and the buffer (p_5) are shared by four workstations. Furthermore, the robot is sequentially shared while the buffer is shared in parallel.

In this model, we have:

Operation place set, $P_A = \{p_{ij}, i = 1, 2, j = 2, 3, 4\}$ where p_{12}, p_{22}, p_{14}, and p_{24} represent machining by four different machines with loading and unloading by the shared robot, respectively and where p_{13} and p_{23} represent the available intermediate parts, stored in the buffer;

Fixed resource place set, $P_B = \{p_5, p_6\}$, $m_0(p_5) = b$, $m_0(p_6) = 1$ where p_5 models the number of empty slots in a buffer, and p_6 models the shared robot; and

Variable resource place set, $P_C = \{p_{11}, p_{21}\}$ where p_{11} and p_{21} model the availability of raw materials, each of which is fixtured with a pallet.

It can be easily verified that (p_E, D) forms an SME where $D = \{(t_{11}, t_{12}), (t_{21}, t_{22})\} \cup \{(t_{13}, t_{14}), (t_{23}, t_{24})\}$. $C(t_{12}, t_{13}) = C(t_{22}, t_{23}) = b$. However, according to Definition 4.4 regarding the effectiveness of a token capacity, both capacities are not *effective* since both C-places links to the B-place p_5.

Therefore, Theorems 4.3 to 4.5 are not applicable.

Intuitively, a designer may split the above buffer into two parts, one with capacity b_1 and the other with b_2, satisfying $b_1 + b_2 = b$, and may model the buffer as two separate buffers by using two non-shared resources. Then, $C(t_{12}, t_{13}) = b_1$, $C(t_{22}, t_{23}) = b_2$, and $C = \min(b_1, b_2)$ and the theorems are applicable. It follows that if $1 \leq m_0(p_{11}) \leq \min(b_1, b_2)$, and $1 \leq m_0(p_{21}) \leq \min(b_1, b_2)$, then the net is live and reversible.

The method may fail if such a division is difficult or impractical. A more rigorous approach is to include such a case in our design theory. To fulfill this objective, two cases are identified and discussed: *independent C-places and dependent C-places.* It is not difficult to combine both cases to obtain more general results.

Case 1: Independent C-places.

For each transition t in T_a^1, there is a unique C-place p such that $EP(p,t)$ exists. Strictly, $\forall t, t' \in T_a^1$, if $\exists p \in P_C, EP(p,t) \neq \emptyset$, and $EP(p,t') \neq \emptyset$, then $t = t'$; and $\forall p, p' \in P_C$, if $\exists t \in T_a^1$, $EP(p,t) \neq \emptyset$, $EP(p',t) \neq \emptyset$, then $p = p'$. We also assume that one shared resource place at most exists between any two PMEs. In other words, $\forall t \in T_b^i$, $t' \in T_b^{i+1}$ for $i \in N_{L-1}$, $|\{p \in P_B : \exists p_{c1}, p_{c2} \in P_C,\ p_{c1} \neq p_{c2},\ EC(p) \cup EP(t,t') \cup EP(p_{c1},p) \cup EP(p_{c2},p) \neq \emptyset\}| = 1$. We change condition C1 into C11:

C11: $\exists m_0(p), p \in P_C$ satisfies the following equations:

1. $m_0(p) \geq \alpha(p) = \max\{\alpha_i(p) : i \in N_{L-1}\}$;
2. $\forall t \in T_b^i$, $t' \in T_a^{i+1}$, $1 \leq i \leq L-1$, if an A-path $EP(t,t')$ exists, $Z^e(t,t')$ is reversible.
3. For $t \in T_b^i$, T_a^{i+1}, if $\exists$ effective $C(t,t')$, then $\beta^i(p,t'')m_0('') \leq C(t,t')$ where t'' is a transition in T_a^i and $EP(t'',t) \neq \emptyset$.
4. If the above $C(t,t')$ is finite and ineffective, then $\exists p_b \in P_B$, t or $t' \in EC(p_b)$. Let $P'' = \{p \in P_C : \exists t \in T_b', \ni EP(p,p_b) \neq \emptyset \text{ in } Z\}$ with $|P''| > 1$. $D_{ab} = \{(t,t') : \exists i \in N_{L-1}, t \in T_b^i,\ t' \in T_a^{i+1}$, t or $t' \in EC(p_b)\}$.

$$cc(t,t') = \begin{cases} 0 & \text{if } C(t,t') \leq m_0(p_b); \\ C(t,t') - m_0(p_b) & \text{otherwise.} \end{cases}$$

Then

$$\sum_{p \in P''} \beta^i(p) m_0(p) \leq m_0(p_b) + \min\{cc(t,t') : (t,t') \in D_{ab}\}.$$

Then, we can prove the following result.

Theorem 4.5 *If C11 and C2 are true, then*

1. *Z' is live if Z is live, and*

2. *Z' is reversible if Z is reversible.*

This theorem has an intuitive explanation: if a type of resources is shared by many PMEs which consume the tokens originating from their own C-places, then these tokens are going to occupy the shared resources. In other words, all tokens from these C-places may take empty slots provided through the token capacity. Therefore, linear summation of initial tokens in these C-places should be limited to that token capacity.

The Petri net in Fig. 4.7 gives a simple example to comment on this theorem. The resulting feasible markings should satisfy:

1. $m_0(p_{11}) \geq 1$, $m_0(p_{21}) \geq 1$; and

2. $m_0(p_{11}) + m_0(p_{21}) \leq b$.

Case 2: Dependent C-places.

Suppose that there is a variable resource place p' whose initial marking is necessary to fire any transition in the SME. Strictly, $\exists p' \in P_C$, if $m_0(p') = 0$ and t is enabled in Z', then $t \notin D$. Such a C-place is called *a principal resource place, or a principal C-place.* The other C-places are dependent on it in the sense that their initial tokens cannot start any operations in the SME without the initial tokens in p'. Note that there may be several principal C-places given a Petri net.

There are two common cases. One occurs when $P_C = \{p'\}$, and the resources in p' are necessary to start any operations in the SME; for example, when there is only one kind of raw materials for final products in a manufacturing system. The second case deals with multiple C-places but one C-place is the principal one. An example is illustrated in the flexible manufacturing system in Chapter 7. Pallets are principal resources since raw blocks and pegs which are modeled as two different C-places need to be fixed with pallets.

Condition C1 may be changed to C12 as follows.

C12: $\exists m_0(p)$, $p \in P_C$ satisfies the following equations:

1. $\forall p \in P_C$, $m_0(p) \geq \alpha(p) = \max\{\alpha^i(p) : i \in N_{L-1}\}$;

2. $\forall t \in T_b^i$, $t' \in T_a^{i+1}$, $1 \leq i \leq L-1$, if an A-path $EP(t,t')$ exists, $Z^e(t,t')$ is reversible.
3. For $t \in T_b^i$ and $t' \in T_a^{i+1}$, if $\beta^i(p') \leq 1$, $m_0(p') \leq C^i(t,t')$, otherwise, $\beta^i(p')m_0(p') \leq C^i(t,t')$.

Note that since all C-places are *dependent*, we do not need to care whether token capacity between two PMEs is effective. In other words, resources in such systems may be shared by different processes, but not by initial tokens in all C-places. The following theorem can be proved.

Theorem 4.6 *If C12 and C2 are true, then*

1. *Z' is live if Z is live, and*

2. *Z' is reversible if Z is reversible.*

Example 4.8: Consider the Petri net shown in Fig. 4.8. It results from minor modification of the net in Fig. 4.7 and the same places of two nets have the same explanation unless otherwise stated. We have:

Operation place set, $P_A = \{p_{ij}, 1 \leq i \leq 2, 2 \leq j \leq 4\} \cup \{p_1\}$ where the explanation of p_{ij} for $1 \leq i \leq 2$, $2 \leq j \leq 4$ is the same as the net shown is Fig. 4.7, and p_1 represents parts being fixtured with pallets;

Fixed resource place set, $P_B = \{p_5, p_6\}$, $m_0(p_5) = b$, $m_0(p_6) = 1$; and

Variable resource place set, $P_C = \{p_{11}, p_{21}\}$ where p_{11} and p_{21} model the availability of raw materials and pallets, respectively.

It is very easy to see that both C-places are *dependent* on each other. In other words, either can be a principal resource place. Practically, absence of either raw materials or pallets starts nothing on this system. Using Theorem 4.6, we find that feasible initial markings satisfy:

$$m_0(p_{11}) \geq 1, \; m_0(p_{21}) \geq 1,$$

and

$$m_0(p_{11}) \leq b, \text{ or } m_0(p_{21}) \leq b.$$

4.7 Generalized Sequential Mutual Exclusions

Parallel mutual exclusions has been generalized to cases with choice operations in Section 3.5. This section will generalize the SME concept to deal with more cases for automated manufacturing systems.

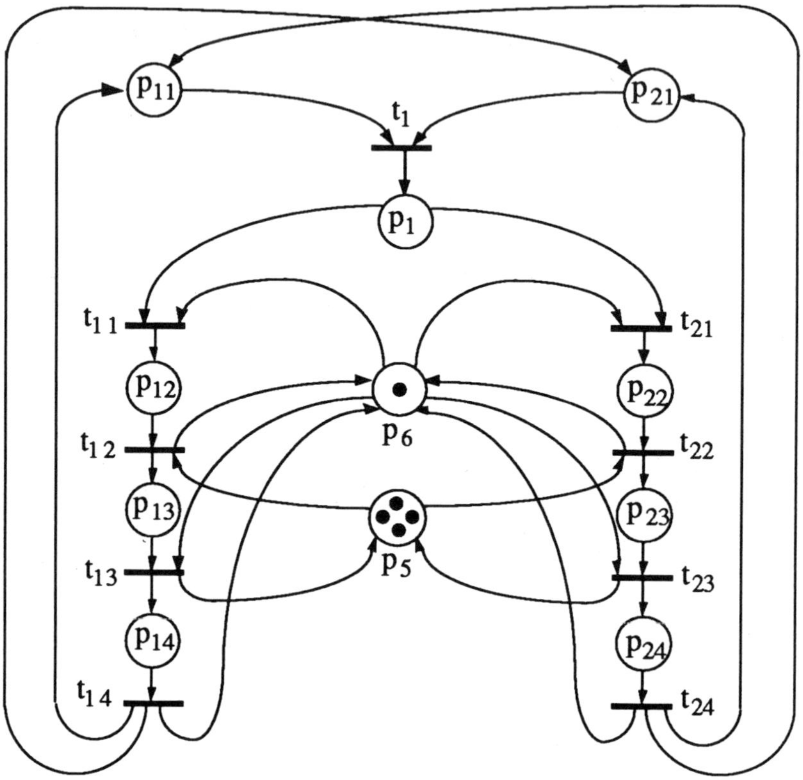

Figure 4.8: A Petri net in Example 4.8

4.7.1 Definition

The following definition deals with *a generalized sequential mutual exclusion* in the sense that two sets of transitions, e.g., T_{a1} and T_{a2}, may be enabled simultaneously only after some transitions in T_{a1} fire from an initial marking at least once.

Definition 4.7 Given a marked Petri net

$$Z' = (P', T', I', O', m_0),$$

$$P' = P_A \cup (P_B \cup \{p_E\}) \cup P_C,$$

a generalized sequential mutual exclusion (GSME) exists if and only if

1. $\exists(p_E, D)$ such that $D = D^1 \cup D^2 \cup \ldots \cup D^L$ where $L \geq 2$, $D^i = \{(T^i_{a1}, T^i_{b1}), (T^i_{a2}, T^i_{b2}), \ldots, (T^i_{ak_i}, T^i_{bk_i})\}$, satisfying
   (a) (p_E, D^i) is a k_i-generalized parallel mutual exclusion in $Z_i = (P, T, I, O, m_0)$ that results from Z with $I(p_E, t) = O(p_E, t) = 0$, $\forall t \in D - D^i$;
   (b) $\forall t \in T^j_a$, $1 \leq j < i$, $\exists t' \in T^i_a$, $\ni$ an A-path $EP(t', t) \neq \emptyset$;
   (c) if an A-path $EP(r, r') \neq \emptyset$ for $r \in D^i$, $r' \in D^j$, then $i < j$; and
   (d) if $EP(t', r') \neq \emptyset$, $t' \in T^i_b$, $r' \in T^j_b$, $i < j$, then $\exists t \in T^j_a$, $\ni t \in EP(t', r')$;
2. $\forall m_0, g_i$, and $i \in N_{L-1}$, if $m_0[g_i>$ enables $t \in T^{i+1}_a$, then $\#(g_i, T^i_a) = \#(g_i, T^i_b) \geq 1$.
3. $\forall m_0$, if $t \in T^i_a$ fires at $m \in R(Z_i, m_0)$, then $\forall g_j$ containing no $t \in T^i_b$, $\exists h_j$ containing no transition in T^j_a, $1 \leq j < i$ if $i > 1$, $\ni m[tg_jh_j>$ enables a transition in T^i_b in Z_i, where $T^i_a = \{T^i_{av}, v \in N_{k_i}\}$, $T^i_b = \{T^i_{bv}, v \in N_{k_i}\}$ for $1 \leq i \leq L$,and $T_a = T^1_a \cup T^2_a \cup \ldots \cup T^L_a$.

In this definition, D^i is also viewed as a set of transitions which are involved in it for convenience, and so is D.

The properties similar to those possessed by an SME can be derived, which appear in Section 4.2. Note that a GSME is degraded as an SME when each of its GPMEs becomes a PME.

Example 4.9: The Petri net in Fig. 4.9 is obtained from Fig. 3.10 in Section 3.5.2 through the addition of arcs, i.e., (p_E, t_4), (p_E, t_5), (t_6, p_E), (p_E, t_9), and (t_{10}, p_E).

$$D^1 = \{(\{t_1\}, \{t_2, t_3\}), (\{t_7\}, \{t_8\})\}, D^2 = \{(\{t_4, t_5\}, \{t_6\}), (\{t_9\}, \{t_{10}\})\}.$$

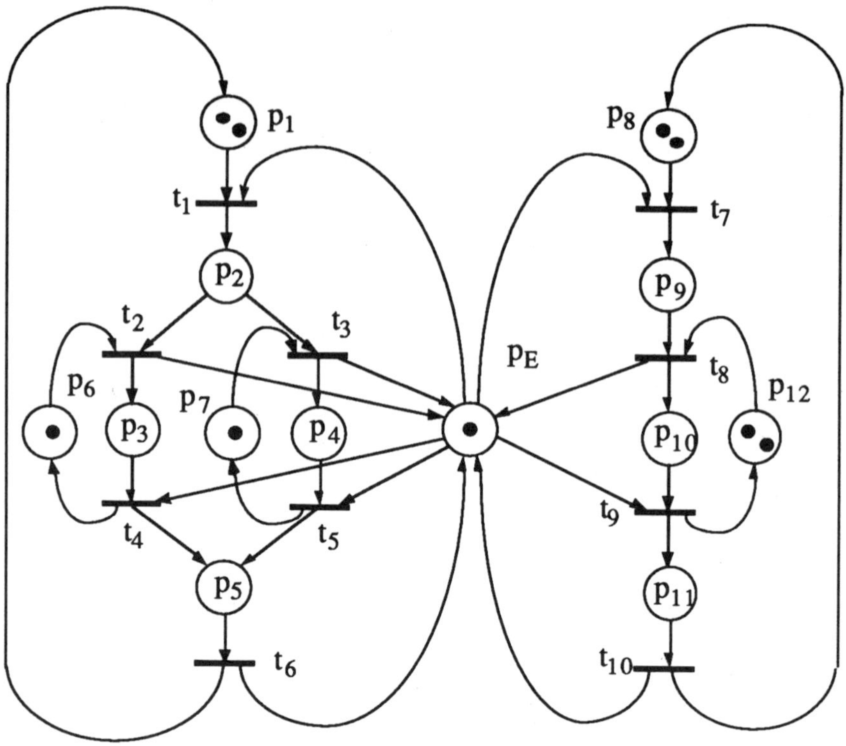

Figure 4.9: A GSME example

The initial marking is

$$m_0 = (m_0(p_1), 0, 0, 0, 0, 1, 1, m_0(p_8), 0, 0, 0, 2, 1)^T.$$

Conditions 1 to 3 can be easily verified. Therefore, $(p_E, D^1 \cup D^2)$ is a GSME.

The liveness and reversibility of the net can be affected if improper initial tokens are distributed in a Petri net which contains a GSME. For the example shown in Fig. 4.9, the system can be blocked when $m_0(p_8) \geq 3$. Specifically, when $m_0(p_8) = 3$, the dead state can be reached by firing $t_7 t_8 t_7 t_8 t_7$; the dead marking for this net is

$$(m_0(p_1), 0, 0, 0, 0, 1, 1, 0, 1, 2, 0, 0, 0)^T.$$

The right side of the net can be viewed as the Petri net model of the system discussed in Example 4.1. When the partial deadlock in the right side is reached, no transition of the net's left side can fire either.

4.7.2 Token Capacity of Generalized Sequential Mutual Exclusions

Given Z' having a GSME, the definitions of three kinds of subnets Z, Z^e, and Z^α remain unchanged. These definitions appear in Section 4.3.2. The definitions of parameters α, β, and γ are given in Section 4.4 and used here without any change.

In order to reveal the characteristics of a GSME in a Petri net, the following discussion is confined to a Petri net under Assumptions 4.2 and 4.3. These assumptions make it easier to calculate the token capacities among different GPMEs of a GSME in a net.

Assumption 4.2 The token capacity between t and t' depends on at most one fixed resource place for $t \in T_b^i$, $t' \in T_a^{i+1}$, $i \in N_{L-1}$ if there is an A-path between them.

Suppose that the token capacity depends on a B-place p. In this case we have

$$C(t, t') = m_0(p).$$

This should be true even when $m_0(p)$ is changing.

The assumption seems restrictive. However, it holds true for most applications. Moreover, there are transformations which can be used to convert some classes of Petri nets into nets which satisfy this assumption.

Assumption 4.3 For any two A-paths $EP(p, p') = ptp_1 \cdots t_1p'$ and $ptp_2 \cdots t_2p'$, either $p_1 = p_2$ or $t_1 = t_2$ in a GSME of Z'.

This assumption states that a token in a place p cannot lead to two tokens in an identical place. A Petri net structure is shown in Fig. 4.10 which cannot satisfy this assumption. Suppose that there is a token in p. The firing of t leads one token to both p_1 and p_2. After t_1 and t_2 fire, place p' obtains two tokens.

Only a few manufacturing examples can result in the structure shown in this figure. For example, a disassemble process can decompose a complete part into several separate parts. Future research should include such cases in the proposed mutual exclusion theory.

Under these two assumptions, the token capacity of a GSME can be easily derived as follows. First, the achievability and effectiveness of token capacities can be extended to the cases we considered. We then incorporate these conditions under which a token capacity is effective and achievable into the

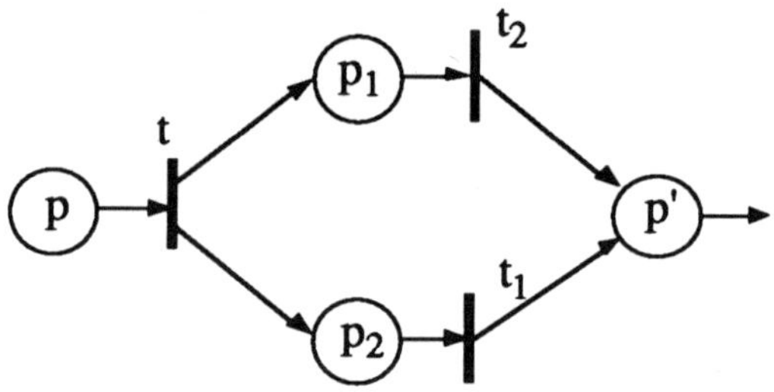

Figure 4.10: A Petri net which does not satisfy Assumption 4.3

definition of token capacity for the GSMEs. This eliminates the need to repeat the definitions of achievability and effectiveness.

Definition 4.8 Suppose that $\forall t \in T^i_{bu}$, $t' \in T^{i+1}_{av}$, $1 \leq u \leq k_i$, $1 \leq v \leq k_{i+1}$, $1 \leq i \leq L-1$, if A-path $EP(t,t') \neq \emptyset$, then $Z^e(t,t')$ is reversible and if $C(t,t') = m_0(p)$, then p links no transitions in $EP(r,r')$ for any $r \in T^j_{bu'}$, $r' \in T^{j+1}_{av'}$, $1 \leq u' \leq k_j$, $1 \leq v' \leq k_{j+1}$, $1 \leq j \leq L-1$, if either $u' \neq u$, $v' \neq v$, or $i \neq j$. The token capacity between D^i and D^{i+1} is defined as: $C^i = \min\{C^i_j, 1 \leq j \leq k_i\}, i \in N_{L-1}$, where

$$C^i_j = \begin{cases} \infty & \text{if } C(t,t') = \infty \;\&\; A\text{-path } EP(t,t') \neq \emptyset \\ & \text{for } t \in T^i_{bj} \;\&\; t' \in T^{i+1}_{av}; \\ \sum_{p \in P(i,j)} m_0(p) & \text{Otherwise.} \end{cases}$$

where

$$P(i,j) = \{p \in P_B : \exists t \in T^i_{bj}, t' \in T^{i+1}_{av}, \ni A\text{-path } EP(t,t') \text{ exists } \}.$$

The conditions in this definition are necessary to allow such a simple calculation. They state that a B-place cannot be shared among different pairs of transition sets of GPMEs. They also require that subnets $Z^e(t,t')$ whose definition is given in Section 4.3.2 are reversible where t is from the ith GPME and t' is from the $(i+1)$st GPME. These conditions, in fact, guarantee the effectiveness of the token capacities.

It should be mentioned that the computation of C^i is very easy because of the above conditions and assumptions. The achievability of the token capacities is ensured through Assumption 4.2.

The calculation of the token capacity and parameters α, β, and γ for a GSME is illustrated in the following simple example. The achievability of the token capacity is also verified.

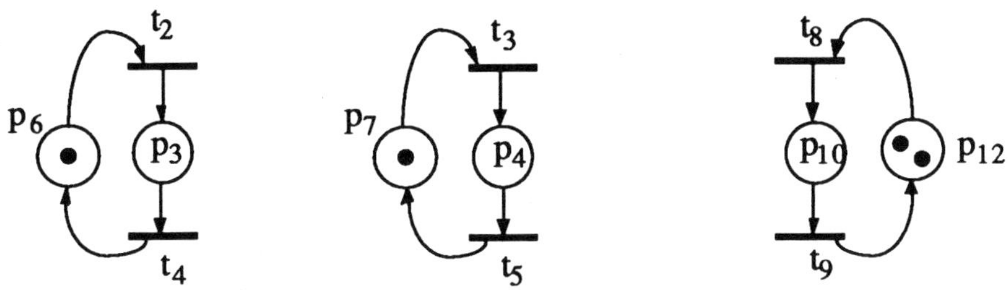

Figure 4.11: Sub-Petri nets of the net in Fig. 4.9

Example 4.10: For the Petri net discussed in the last example in Fig. 4.9, C^1 can be computed as follows. First, we have:

$$D^1 = \{(T^1_{a1}, T^1_{b1}), (T^1_{a2}, T^1_{b2})\}, D^2 = \{(T^2_{a1}, T^2_{b1}), (T^2_{a2}, T^2_{b2})$$

where

$$T^1_{a1} = \{t_1\}, T^1_{b1} = \{t_2, t_3\}, T^1_{a2} = \{t_7\}, T^1_{b2} = \{t_8\},$$

and

$$T^2_{a1} = \{t_4, t_5\}, T^2_{b1} = \{t_6\}, T^2_{a2} = \{t_9\}, T^2_{b2} = \{t_{10}\}.$$

We can easily verify that Assumptions 4.2 and 4.3 and the conditions in Definition 3.2 in Section 3.5 are satisfied. For example, all nets of Z^e are reversible as shown in Fig. 4.11. Therefore, we have:

$$C^1_1 = m_0(p_6) + m_0(p_7) = 2$$

$$C^1_2 = m_0(p_{12}) = 2$$

$$C^1 = \min(C^1_1, C^1_2) = 2$$

Next, from the net Z shown in Fig. 4.12, C^1 is achievable because t_1 is enabled if $m_0(p_1) > 2$ after t_2 and t_3 fire once without firing t_4 and t_5 respectively. Similarly, since $C^1 = C(t_8, t_9) = 2$, t_7 is enabled after t_8 fires twice without firing t_9 when $m_0(p_8) > 2$.

Finally, from the net Z^α shown in Fig. 4.13, we obtain:

$$\beta(p_1, t_1) = 1, \beta(p_1, t_7) = 0, \beta(p_1) = \max\{\beta(p_1, t_1), \beta(p_1, t_7)\} = 1;$$

$$\alpha(p_1, t_1) = 1, \alpha(p_1, t_7) = 0, \alpha(p_1) = \max\{\alpha(p_1, t_1), \alpha(p_1, t_7)\} = 1;$$

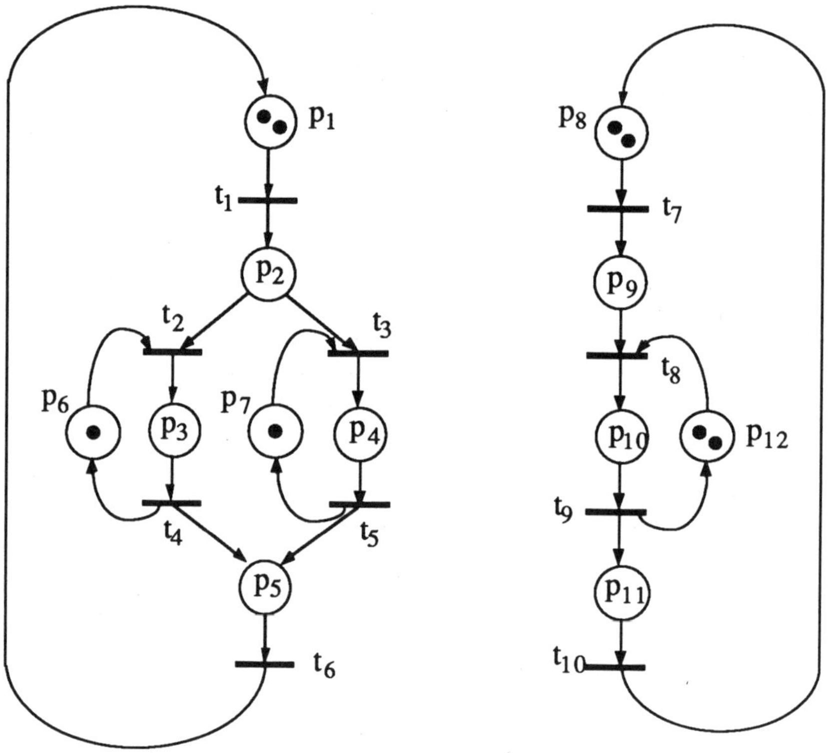

Figure 4.12: Z and verification of the achievable C^1

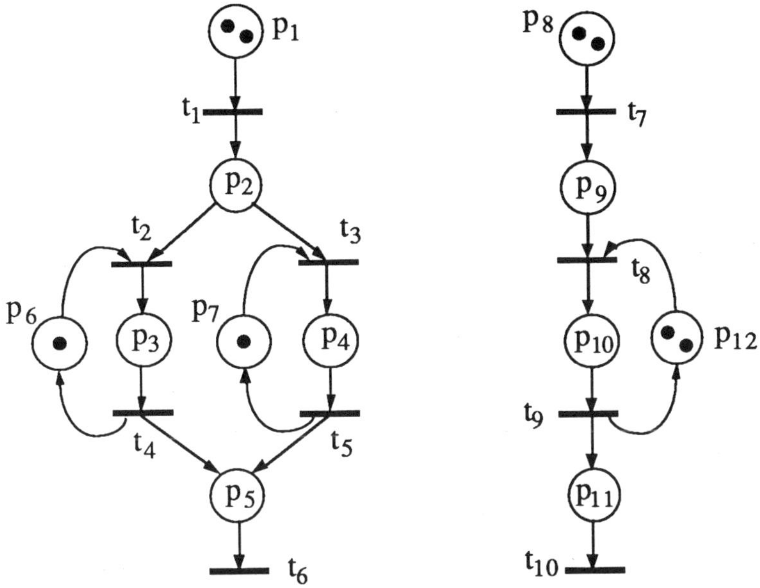

Figure 4.13: Z_α and calculation of α, β and, γ

Similarly,

$$\beta(p_8) = 1, \alpha(p_8) = 1.$$

Obviously,

$$\gamma(p_1) = \gamma(p_8) = \infty.$$

4.7.3 Preservation of System Properties

Suppose that a Petri net Z' satisfies Assumptions 2.1 to 2.2, and Assumptions 4.2 and 4.3. In addition,

1. $Z = (P_A \cup P_B \cup P_C, T, I, O, m_0)$ is a subnet of $Z' = (P', T', I', O', m_0')$, where $P' = P_A \cup (P_B \cup \{p_E\}) \cup P_C$ and $T' = T$;

2. $D = D^1 \cup D^2 \cup \ldots \cup D^L$, $k_i = |D^i| \geq 1, L \geq 2$, where $D^i = \{(T_{a1}^i, T_{b1}^i), (T_{a2}^i, T_{b1}^i), \ldots, (T_{ak_i}^i, T_{bk_i}^i)\}$, $i \in N_L$; and (p_E, D) is a GSME;

Theorem 4.7 *(Generalized sequential mutual exclusion) Suppose Z and Z′ are specified as in the preceding paragraph.*

1. *Z' is bounded or safe if Z is bounded or safe; if either*

   (a) *In Z, $\exists C^i$, m_0, α, and β, $\ni m_0(p) \geq \alpha(p) = \max\{\alpha^i(p) : i \in N_{L-1}\}$, and $\beta^i(p)m_0(p) \leq C^i$, $p \in P_C$; or*

   (b) *There is a principal resource place p' in P_C, and $\exists m_0$, α, and β, $\ni m_0(p) \geq \alpha(p) = \max\{\alpha^i(p) : i \in N_{L-1}\}$, $\forall p \in P_C$, and $\forall t'' \in T^i_{au}$, $t \in T^i_{bu}$, $t' \in T^{i+1}_{av}$, $1 \leq u \leq k_i$, $1 \leq v \leq k_{i+1}$, $1 \leq i \leq L-1$, if A-paths $EP(t,t') \neq \emptyset$, $EP(t'',t) \neq \emptyset$, and $Z^e(t,t')$ is reversible,*

   $$\beta^i(p',t'')m_0(p') \leq C(t,t'),$$

   then

2. *Z' is live if Z is live;*

3. *Z' is reversible if Z is reversible.*

Note that this theorem can be proved using the same idea that is used to prove Theorems 4.2 to 4.4.

Theoretical results can be developed for more complicated cases where token capacities may become ineffective and unachievable.

The simple application of this theorem can be directed to the Petri net shown in Fig. 4.9 in Example 4.9. With no consideration of the shared resource, the net can be easily proved to be bounded, live, and reversible under any initial markings:

$$m_0(p_1) > 0 \text{ and } m_0(p_8) > 0.$$

Using Theorem 4.7, we obtain the following results by using the token capacity and parameters computed in Example 4.10:

The net is bounded, live, and reversible under the initial markings:

$$0 < m_0(p_1) \leq 2 \text{ and } 0 < m_0(p_8) \leq 2.$$

A more complicated manufacturing example is modeled and discussed in Chapter 6 using this theory. Other examples can also be found in [Zhou 90a, 90f, 91d].

4.8 Comparison With Previous Research

4.8.1 Mutual Exclusion Concepts in the Literature

The literature provides some definitions [Agerwala 79; Silva 85a] and explanations [Peterson 81] about mutual exclusion phenomena. In computer systems, mutual exclusion solutions are used to ensure that only one process at a time is able to use a shared resource or modify shared information [Dijkstra 68; Jacob 89]. However, the following discussion is confined to mutual exclusion concepts in the context of Petri nets.

Mutual exclusion of a set of transitions can be defined using *p*-invariants. According to Agerwala [1979], two transitions are mutually exclusive if they are transitions of the same invariant and the initial marking is such that the sum of the tokens in the places of the invariant is 1. Here a transition t is said to be a transition of an invariant if an input or output place of t is contained in the invariant. For the example in Fig. 2.2(b), any two of the transitions t_1, t_2, t_3, and t_4 are *mutually exclusive* since they are in the invariant $\{p_3, p_4, p_5\}$ and $m_0(p_3) + m_0(p_4) + m_0(p_5) = 1$ no matter whether initial tokens exist in other places or not. Similarly, for the example in Fig. 4.2, it can be easily verified that any two transitions of t_1, t_2, t_3, and t_4 are mutually exclusive. However, the two nets behave very differently as discussed earlier. It can be easily proved that any two of the transitions in a PME or SME defined as before are mutually exclusive according to Agerwala's definition.

Silva [1985a] formulates a different definition of *mutual exclusion* for two places, not transitions. Two places have mutual exclusion if they cannot be simultaneously marked for a given initial marking [Silva 85a]. This is a broad definition of mutual exclusions in the sense that many pairs of places may have mutual exclusion. For the example shown in Fig. 2.2(a), not only do any pair of p_3, p_4, and p_5 have mutual exclusion, but also p_1 and p_3, and p_2 and p_4 have mutual exclusion with respect to $m_0 = (1, 1, 0, 0, 1)^{\tau}$. When $m_0 = (2, 2, 0, 0,)^{\tau}$, any pair of p_3, p_4, and p_5 still have mutual exclusion, but p_1 and p_3, and p_2 and p_4 do not. In this respect, this definition differs from those proposed in this book.

In summary, our definitions of mutual exclusion are motivated to serve the purpose of Petri net synthesis for automated manufacturing systems.

4.8.2 Resource Allocation Problems

Dijkstra [1968] gave the Banker's Problem as an example of a resource-sharing system. A banker wishes to share a constant capital of florins among a fixed number of customers. Each customer has limits on its maximal claim of florins, borrows the florins one by one from the banker, and later returns the loan money after the maximal claim is reached. The problem for the banker is to prevent deadlock situations where neither the banker has any florins himself nor any customer has enough money to return his credit. Hauschildt and Valk [1985] have modeled such a problem using Petri nets and derived the set of all safe states. Their results are also used to solve the multidimensional Banker's Problem.

Several resource allocation problems in computer science can be viewed as originating from this problem. Datta *et al.* [1984b] have studied deadlock avoidance techniques for the shared disk problem in which Dijkstra's customers are characterized by computers, capital of florins by total disk capacity, and maximum florins by the claim of each computer. When a computer executes a program, it claims the required disk blocks, acquires them one by one, uses them for a finite time period, and releases all claimed disk blocks one by one. In such a system, a deadlock may result from an inappropriate allocation of some critical disk blocks and no program can be complete. Using Petri nets, Datta et al. [1984b] have modeled the system and invented two kinds of control mechanisms to avoid any deadlock.

A similar resource allocation problem was presented by Lautenbach and Thiagarajan [1979]. Sequentially related processes compete for a number of resources which are stored in a central depository. The demands for resources by a process are known at each step. All the seizures of a resource type have occurred before the release of resources takes place. A strategy is formulated to guarantee the liveness of the net by granting or denying a particular request based on an initial computation.

This problem has been generalized along two different lines. Suraj [1980] has allowed the system to have several cyclic processes. Between every two processes, a control place marked initially with tokens is designed and added to the original net such that the system is deadlock-free. All and only those safe states in the original net are preserved in the designed net. Generalizing the problem by allowing a used resource to be released at any time, Tazza [1987] has succeeded in avoiding any deadlock and defining an allocation method corresponding to a relative maximum for the throughput of the system.

One common point is that all these problems have only one type of shared resources, for example, florins or disk blocks. These identical shared resources may be numerous. Hence, commonly understood mutual exclusion may not be involved, for example, the example discussed by Suraj [1980]. Differing from the resource-sharing assumption in their problems, the shared resource considered in our PME and SME is single, and corresponds to different active components, such as a loading/unloading robot shared by several machines in a manufacturing system. Also, many distinct shared resources may be present in our concerned systems.

However, PME and SME may be further developed to accommodate different resource-sharing requirements. Along this direction, a buffer later becomes of concern in the next chapter. A buffer may be viewed as a composition of many identical resources and is often shared by different processes.

Other related work includes performance evaluation of systems of cyclic sequential processes with mutual exclusion [Magott 85, 87]. The problem of finding the cycle time of a deterministic time system is investigated for safe, live, and reversible Petri nets. Self-loops are used to model a single shared resource. We find that the examples presented are live and reversible no matter how many tokens are initially distributed in Magott's systems.

Summarizing, this chapter has presented the concepts and results for Petri net modeling of resources which are shared by sequentially related processes. They have been illustrated through several simple manufacturing examples. The results will be used a basis for bottom up design of shared resources in the next chapter.

Chapter 5

Hybrid Synthesis of Petri Nets

5.1 Introduction

5.1.1 Motivation and Objectives

For linear, continuous, or synchronous discrete systems, there exist methodologies for modeling and control system design, which, if applied correctly, will provide desirable performance properties [Zhou 92a]. However, for discrete event dynamic systems, effective design for real-world applications theory is lacking for real-world applications. Petri nets can be developed to fill this need. They have been used to model discrete event dynamic systems such as automated manufacturing systems and computer systems. These can also lead to controllers as we discussed in Section 1.3. In addition, there exist validation, simulation, and performance analysis tools for Petri nets. They are useful for designing Petri net controllers for manufacturing systems.

Problems arise when the complexity of a real-world system leads to a large Petri net which has many places and transitions. It is possible to construct such a large net by, for example, merging subnets. However, it is nearly impossible to analyze it globally using the reachability graph or the invariant methods. For example, a Petri net modeling a factory producing corrugated fiberboard has 92 places, 59 transitions, and 174 arcs [Lee 85]. The authors emphatically comment on the extreme complexity of the structural analysis for this model [Lee 85]. Hence, many researchers have developed many reduction techniques which can serve as synthesis methods if appropriately applied. In another example [Kasturia 88], a Petri net model for a two-workstation manufacturing system was analyzed using the invariant method. The incidence matrix has 58 rows and 42 columns. Forty equations are obtained for the p-invariants and sixty-one for t-invariants. One can imagine the complexity when so many equations are used to analyze liveness and boundedness. The reachability method was not feasible for this system because of the state explosion problem.

Also, an early design error can lead to a wrong net model whose properties cannot be verified [Valette 87]. To find such an error is very difficult, and after finding the error, modifying the design and re-analyzing the model might require a tremendous amount of work.

Finally, high-level Petri nets, such as colored Petri nets [Jensen 81] or some extended Petri nets [Gentina 87], could be used to overcome the problem of graphical complexity. However, these lack the analysis methods and software tools which are available for ordinary Petri nets.

All these experiences urge us to study a systematic method for synthesis of Petri nets for manufacturing systems. This methodology addresses this problem in two ways. First, the design process is divided into two major stages: (a) in a top-down manner, a designer starts with an aggregate Petri net description and uses stepwise refinement of operation places and/or transitions to include more detail until the desired level is achieved; (b) the resource places are added to the net in a bottom-up way. If needed, (a) and (b) could be exercised alternately. Thus the complexity of the detailed problem may be alleviated. Second, in order to circumvent the qualitative analysis problem for a complex system, the methodology includes a set of basic design modules and mutual exclusion structures which can be used in the proposed synthesis procedure. Both the net structure and the initial marking are designed so that the synthesized Petri nets are guaranteed to have desirable qualitative properties such as boundedness, liveness, and reversibility. Therefore, a general problem for modeling a manufacturing system using a Petri net can be stated as follows:

> *Given the specifications of a manufacturing system, model the system as a Petri net such that its structure and initial markings make it bounded, live, and reversible.*

5.1.2 Organization of This Chapter

This chapter is organized as follows. Section 5.2 discusses the design of a first-level Petri net as an aggregate model for different manufacturing systems. Two common structures are given: choice-free Petri net structures and choice-synchronization Petri net structures. This is important in the use of the proposed methodology. Design of aggregate net models is also crucial to apply other top-down design methods [Valette 79; Suzuki 83; Zhou 89b].

Section 5.3 introduces formally four basic Petri net design modules: sequence, parallel, choice, and decision-free choice. The refinement theory [Valette 79; Suzuki 80, 83a] is summarized and is extended to include reversibility of Petri nets.

Section 5.4 provides several design methods for both a simple and generalized buffers using Petri nets. Two theorems about maintenance of system properties are presented and proved when the proposed buffer modules are included in Petri nets. The application scope for such theorems is by imposing some conditions on the original Petri nets.

Section 5.5 proposes a hybrid Petri net synthesis procedure. The design of Petri nets for a system is divided into two major parts: top-down design and bottom-up design. In the first stage, all operation places are refined and non-shared resources are added. In the second stage, buffer modules and resource-sharing problems are taken into account. The mutual exclusion theory developed in Chapters 3 and 4 is applied so that the qualitative properties of Petri nets are preserved.

The next two chapters will illustrate the methodology by Petri net modeling and control of manufacturing systems.

5.2 Design of First-level Petri Net Models

Design of the first-level Petri net models for different manufacturing systems is so important in top-down approaches that it is treated as a separate topic discussed in this section. The following presents two basic kinds of net models which are commonly used.

5.2.1 Choice-free Petri Nets

Most manufacturing systems produce different kinds of parts from different raw materials and/or from different entries. These parts may be used for assembly in other systems which are designed separately. Theoretically, design of Petri nets for such systems can begin with only a single transition as shown in Fig. 5.1(a). It can be converted into a Petri net having a single place and a single transition as shown in Fig. 5.1(b). Such a transformation can be proved to preserve the properties if p_1 has some initial tokens. It can be further augmented to the net in Fig. 5.1(c) which models a system with a single resource and a single operation. The system starts the operation when the resource is available. When several major resources are needed to start a system with several concurrent operations, the extension of the above net can be depicted as in Fig. 5.1(d). The matrix expression of the net follows:

$$Z = (P_A \cup P_B \cup P_C, T, I, O, m_0)$$

where

$$P_A = \{p_{n+1}, ..., p_{n+k}\}$$

$$P_B = \emptyset$$

$$P_C = \{p_1, p_2, ..., p_n\}$$

$$T = \{t_1, t_2\}$$

$$I = \begin{pmatrix} 1_n & 0_n \\ 0_k & 1_k \end{pmatrix}$$

and

$$O = \begin{pmatrix} 0_n & 1_n \\ 1_k & 0_k \end{pmatrix}$$

Here 0_k is a zero k-dimension vector and 1_k is a k-dimension vector with each element being 1.

This net can easily be shown to be K-bounded, live, and, reversible with respect to the initial marking: $m_0(p_i) > 0$ for $1 \leq i \leq n$, where $K = \max\{m_0(p_i), 1 \leq i \leq n\}$ The meaning of the net can be interpreted as a system where k parallel operations will start when n resources are available.

When a system can be decomposed into several independent systems, several of the above nets may be used as a Petri net model for the entire system. The manufacturing example will be given later in Chapter 6, where a first-level model is a combination of two separate Petri nets shown in Fig. 5.1(c).

5.2.2 Choice-Synchronization Petri net

When the system involves assembly tasks and separate parts are produced from the same raw materials at the same entry point, then the aggregate Petri net should contain a so-called choice-synchronization structure. To illustrate this concept, we show two typical choice-synchronization structures in Fig. 5.2(a) and (b). In Fig. 5.2(a), when a token resides in place p_1, either t_1 can fire to mark p_2 or t_2 can fire to mark p_3 with a token. When there are tokens in both places p_2 and p_3, transition t_3 is enabled. If there is no restriction on the number of times t_1 or t_2 can fire, then the net may be partially deadlocked since either p_2 or p_3 may never hold a token and thus t_3 may not be enabled. A strict firing order for t_1 and t_2 may be implemented by designing a pair of places called control places, p_4 and p_5, with initial marking

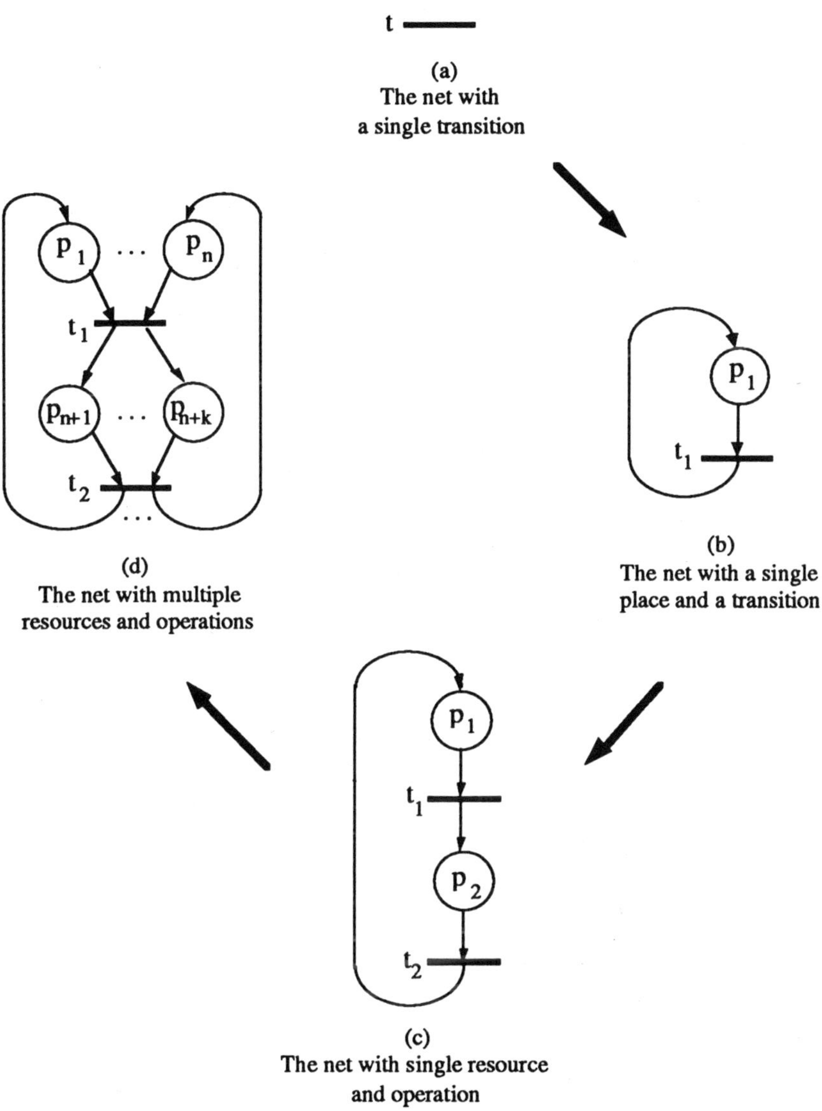

Figure 5.1: Four aggregate Petri net models

as shown in Fig. 5.2(b). In this structure t_1 and t_2 will fire alternately with t_1 firing first.

Therefore, we can design the first aggregate Petri net model as shown in Fig. 5.2(c). This net is, in fact, a generalized Petri net because there are two arcs from t_3 to p_1, i.e., $O(p_1, t_3) = 2$. However, it can be easily changed to an ordinary Petri net by inserting an extra place and transition between t_3 and p_1. It should be noted that two control places p_4 and p_5 are introduced to guarantee that the transition t_1 is enabled first, and that t_1 and t_2 fire alternatively when the initial marking is $(m_0(p_1), 0, 0, 1, 0)^\tau$ and $m_0(p_1) \geq 2$. This Petri net can be proved to be bounded, live, and reversible with the above initial marking. Without the control places, the net would contain a system deadlock.

For some manufacturing systems, the finished parts have to follow a particular order to enter an assembly station: then, the net model can be designed as shown in Fig. 5.2(d), where the first processed part needs to enter the assembly station first. Place p_6 may be interpreted as the availability of an assembly station, and p_7 as the availability of the first part in the station. These Petri nets can be easily modified to satisfy different specifications. The Petri net shown in Fig. 5.2(d) can be also proved to be bounded, live, and reversible when $m_0 = (m_0(p_1), 0, 0, 1, 0)^\tau$ with $m_0(p_1) \geq 2$. The matrix expressions of the net depicted in Fig. 5.2 (b) is given as follows:

$$Z = (P_A \cup P_B \cup P_C, T, I, O, m_0)$$

where

$$P_A = \{p_2, p_3, p_5, p_7\}$$

$$P_B = \{p_4, p_6\}$$

$$P_C = \{p_1\}$$

$$T = \{t_1, t_2, t_3, t_4\}$$

$$I = \begin{pmatrix} 1 & 1 & 0 & 0 \\ 0 & 0 & 1 & 0 \\ 0 & 0 & 0 & 1 \\ 1 & 0 & 0 & 0 \\ 0 & 1 & 0 & 0 \\ 0 & 0 & 1 & 0 \\ 0 & 0 & 0 & 1 \end{pmatrix}$$

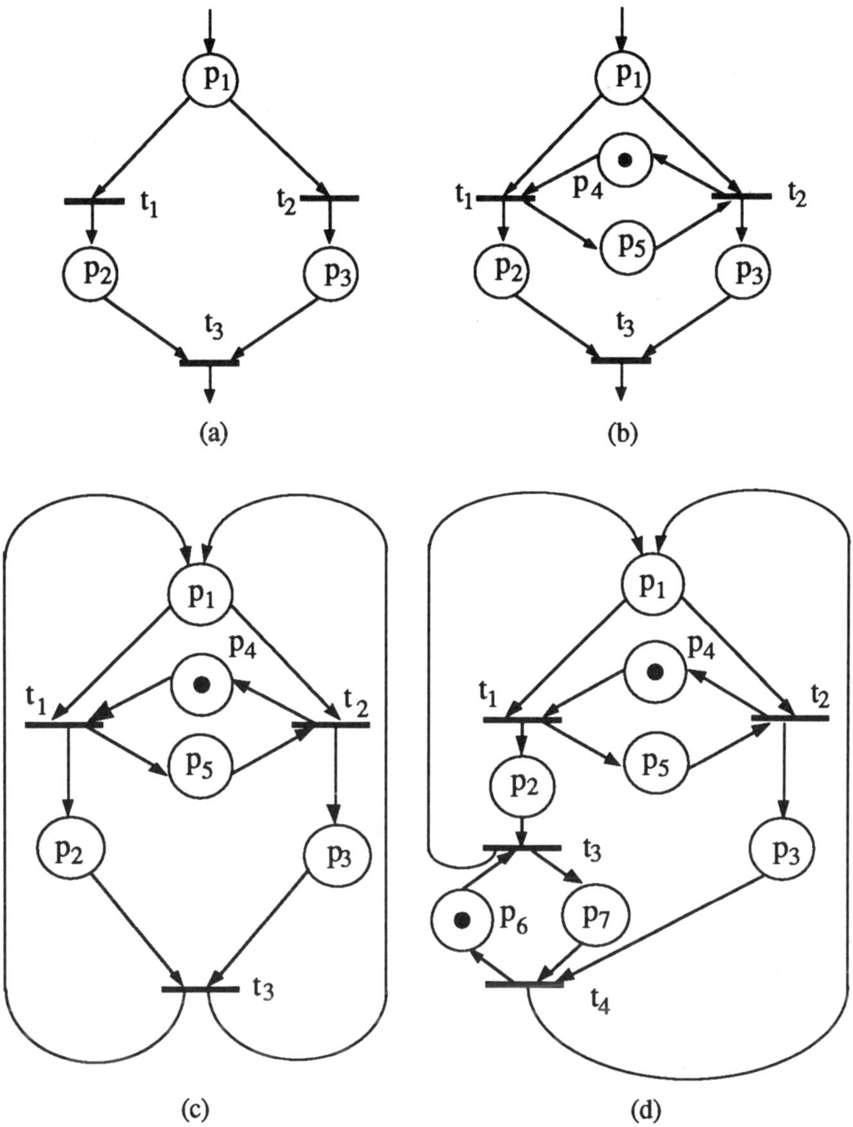

Figure 5.2: Choice-synchronization structures: (a) structure with no restriction on t_1 and t_2, (b) with fixed sequence on t_1 and t_2, (c) aggregate net with fixed sequence on t_1 and t_2, and (d) aggregate net with fixed sequence on t_1 and t_2, and on t_3 and t_4

and

$$O = \begin{pmatrix} 0 & 0 & 1 & 1 \\ 1 & 0 & 0 & 0 \\ 0 & 1 & 0 & 0 \\ 0 & 1 & 0 & 0 \\ 1 & 0 & 0 & 0 \\ 0 & 0 & 0 & 1 \\ 0 & 0 & 1 & 0 \end{pmatrix}$$

Two kinds of variable resources in a system may exist and need to be fixtured together for any further processing or for transportation purposes. In order to deal with this case, the above net in Fig. 5.2(b) can be modified as shown in Fig. 5.3. For example, the case presented can be understood as a system which requires a raw material be fixtured onto a pallet for transportation and/or handling.

$$P_C = \{p_{c1}, p_{c2}\}$$

Place p_1 in Fig. 5.3 may model the operation: fixturing raw material onto a pallet.

This net is used in Chapter 7 as the basis for the aggregate model of a flexible manufacturing system.

The extensions of this model to deal with k ($k > 2$) possible choices and production of a final product which is assembled from k separate parts can easily be made, and the corresponding nets and initial markings can be designed such that the first nets are bounded, live, and reversible. This exercise is left to the reader.

Also, two kinds of basic structures discussed above may be combined to design a first-level Petri net for modeling and discrete event control of a complicated system.

5.3 Basic Design Modules

The concepts of well-formed blocks proposed by Valette [1979], or well-behaved Petri nets by Suzuki and Murata [1980; 1983a] are very useful in the context of Petri net theory and applications. In our methodology, instead of such general Petri net design modules, four basic design modules are introduced to aid in Petri net synthesis for manufacturing systems. The advantage of these modules is evident: they are specific and, therefore, easily accepted and applied when a practical manufacturing synthesis problem is encountered. Similar work was also done by Bruno and Altman [1971] in designing well-formed control network graphs using their basic control modules.

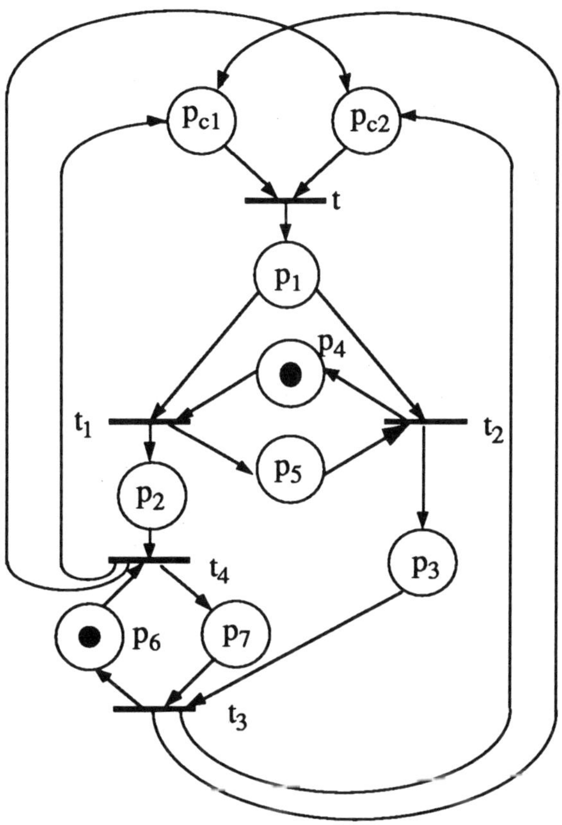

Figure 5.3: An extension of the Petri net shown in Fig. 5.2(d)

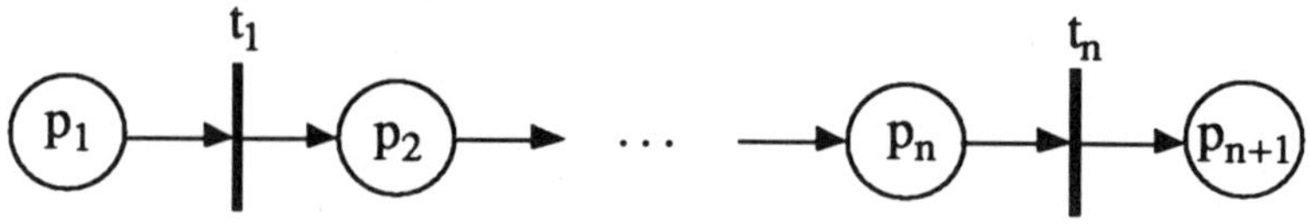

Figure 5.4: Definition of a sequence PN ($n > 0$)

5.3.1 Definitions of Basic Design Modules

Let $S_P = (P, T, I, O, m_{s0})$, where $P = \{p_{in}, p_{out}\} \cup P_s$, and p_{in} and p_{out} are two distinct places. S_P is called *a place PN block*. $S_T = (P, T, I, O, m_{s0})$ is called *a transition PN block* if $T = \{t_{in}, t_{out}\} \cup T_s$, where t_{in} and t_{out} are two distinct transitions.

Definition 5.1 *A sequence PN* module is defined as a Petri net (Fig. 5.4) which consists of $n+1$ places and n transitions ($n > 0$) with an initial marking of zero. Its place PN block $S_P = (\{p_1, p_2, ..., p_{n+1}\}, \{t_1, ..., t_n\}, I, O, 0_{n+1})$ where 0_{n+1} is an $(n+1)$-dimension zero vector with

$$p_{in} = p_1, p_{out} = p_{n+1},$$

$$I = \begin{pmatrix} I_n \\ 0_n^T \end{pmatrix}$$

and

$$O = \begin{pmatrix} 0_n^T \\ I_n \end{pmatrix}$$

where I_n is an $n \times n$ unit matrix.

Its transition PN block

$$S_T = (\{p_2, ..., p_n\}, \{t_1, ..., t_n\}, I, O, 0_{n-1})$$

with

$$t_{in} = t_1, t_{out} = t_n, n > 1$$

$$I = \begin{pmatrix} 0_{n-1} & I_{n-1} \end{pmatrix}$$

and

$$O = \begin{pmatrix} I_{n-1} & 0_{n-1} \end{pmatrix}.$$

In the context of manufacturing systems, a sequence PN represents a series of successive operations. For example, loading Part 1 to Machine 1, processing Part 1 by Machine 1, and unloading Part 1 from Machine 1 are such a

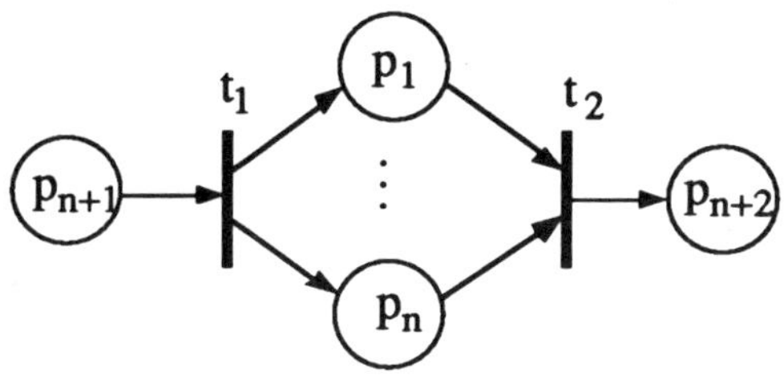

Figure 5.5: Definition of a parallel PN ($n > 1$)

sequence. An operation might be further detailed as a sequence of suboperations, e.g., processing on Machine 1 detailed as loading Program 1, machining using Tool 1, changing to Tool 2, and machining using Tool 2.

Definition 5.2 *A parallel PN* module is defined as a Petri net (Fig. 5.5) which consists of two transitions, $n+2$ places with n parallel places ($n > 1$), and with an initial marking of zero. Its place PN block $S_P = (\{p_1, p_2, ..., p_{n+2}\}, \{t_1, t_2\}, I, O, 0_{n+2})$ with

$$p_{in} = p_{n+1}, p_{out} = p_{n+2},$$

$$I = \begin{pmatrix} 0_n & 1_n \\ 1 & 0 \\ 0 & 0 \end{pmatrix}$$

and

$$O = \begin{pmatrix} 1_n & 0_n \\ 0 & 0 \\ 0 & 1 \end{pmatrix}$$

where 1_n is an n-dimension vector with each element being one.

Its transition PN block $S_T = (\{p_1, p_2, ..., p_n\}, \{t_1, t_2\}, I, O, 0_n)$ with $t_{in} = t_1$, $t_{out} = t_2$, $n > 1$

$$I = \begin{pmatrix} 0_n & I_n \end{pmatrix}$$

and

$$O = \begin{pmatrix} I_n & 0_n \end{pmatrix}.$$

A parallel PN represents n operations which begin execution at the same time. After these n operations terminate, other operations can start. These n operations may end at different times.

Definition 5.3 *A choice PN* module is defined as a Petri net (Fig. 5.6) which consists of $2(n+1)$ transitions and $n+2$ places with n parallel paths

($n > 1$), and with an initial marking of zero. Its place PN block $S_P = (\{p_1, p_2, ..., p_{n+2}\}, \{t_1, ..., t_{2n}\}, I, O, 0_{n+2})$ with

$$p_{in} = p_{n+1}, p_{out} = p_{n+2},$$

$$I = \begin{pmatrix} 0_{n\times n} & I_n \\ 1_n^T & 0_n^T \\ 0_n^T & 0_n^T \end{pmatrix}$$

and

$$O = \begin{pmatrix} I_n & 0_{n\times n} \\ 0_n^T & 0_n^T \\ 0_n^T & 1_n^T \end{pmatrix}$$

where $0_{n\times n}$ is an $n \times n$ zero matrix.

Its transition PN block

$$S_T = (\{p_1, p_2, ..., p_{n+2}\}, \{t_1, t_2, ..., t_{2n+2}\}, I, O, 0_{n+2})$$

with

$$t_{in} = t_{2n+1}, t_{out} = t_{2n+2}, n > 1$$

$$I = \begin{pmatrix} 0_{n\times n} & I_n & 0_n^T & 0_n^T \\ 1_n^T & 0_n^T & 0 & 0 \\ 0_n^T & 1_n^T & 0 & 1 \end{pmatrix}$$

and

$$O = \begin{pmatrix} I_n & 0_{n\times n} & 0_n & 0_n \\ 0_n^T & 0_n^T & 1 & 0 \\ 0_n^T & 1_n^T & 0 & 0 \end{pmatrix}.$$

A choice PN represents n choices for a successive operation. It designates situations in which a given condition or conditions enable multiple competing transitions, or in which an operation can be immediately followed by the choice of one of several other operations. There are other ways of modeling conflict, but not all of them have meaning in the real-world [Freedman 88].

Definition 5.4 *A decision-free choice PN* module is defined as a Petri net (Fig. 5.7) which consists of $2(n+1)$ transitions and $2(n+1)$ places with n parallel paths ($n > 1$), and with an initial marking $m_0 = (0, .., 0, 1, 0, ..., 0)^T$ where its $(n+3)$rd component equals one. Its place PN block

$$S_P = (\{p_1, p_2, ..., p_{2n+2}\}, \{t_1, ..., t_{2n}\}, I, O, \begin{pmatrix} 0_{n+2} \\ 1 \\ 0_{n-1} \end{pmatrix})$$

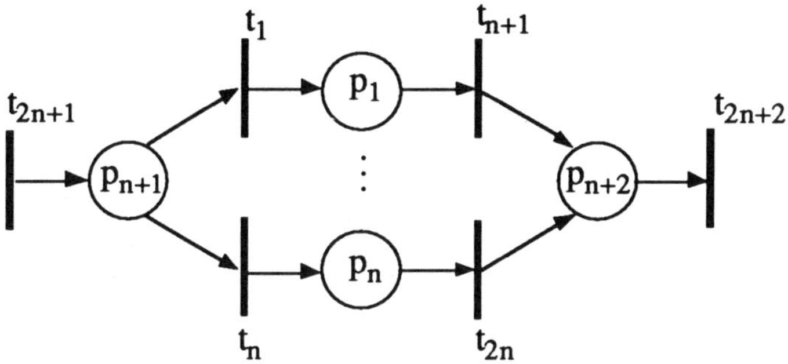

Figure 5.6: Definition of a choice PN ($n > 1$)

with

$$p_{in} = p_{n+1}, p_{out} = p_{n+2},$$

$$I = \begin{pmatrix} 0_{n\times n} & I_n \\ 1_n^T & 0_n^T \\ 0_n^T & 0_n^T \\ I_n^T & 0_{n\times n} \end{pmatrix}$$

and

$$O = \begin{pmatrix} I_n & 0_{n\times n} \\ 0_n^T & 0_n^T \\ 0_n^T & 1_n^T \\ J_n & 0_{n\times n} \end{pmatrix}$$

where $J_n = \begin{pmatrix} 0_{n-1}^T & 1 \\ I_{n-1} & 0_{n-1} \end{pmatrix}$ and I_n is $n \times n$ unit matrix.

Its transition PN block

$$S_T = (\{p_1, p_2, ..., p_{2n+2}\}, \{t_1, t_2, ..., t_{2n+2}\}, I, O, \begin{pmatrix} 0_{n+2} \\ 1 \\ 0_{n-1} \end{pmatrix})$$

with

$$t_{in} = t_{2n+1}, t_{out} = t_{2n+2},$$

$$I = \begin{pmatrix} 0_{n\times n} & I_n & 0_n & 0_n \\ 1_n^T & 0_n^T & 0 & 0 \\ 0_n^T & 0_n^T & 0 & 1 \\ I_n & 0_{n\times n} & 0 & 0 \end{pmatrix}$$

and

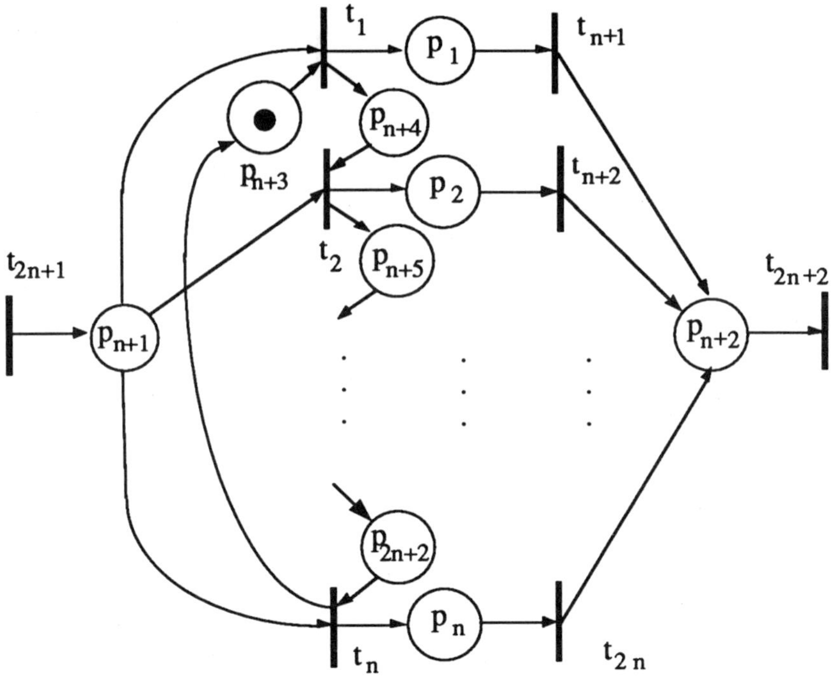

Figure 5.7: Definition of a decision-free choice PN ($n > 1$)

$$O = \begin{pmatrix} I_n & 0_{n\times n} & 0_n & 0_n \\ 0_n^T & 0_n^T & 1 & 0 \\ 0_n^T & 1_n^T & 0 & 0 \\ J_n & 0_{n\times n} & 0 & 0 \end{pmatrix}$$

A decision-free choice PN models n choices for a successive operation with a strict order among them depending on the initial markings. When the initial marking is as shown in Fig. 5.7, then the block enables only t_1 when p_{n+1} gets any token. After t_1 fires and p_{n+1} obtains any token, then only t_2 is enabled, and so on. After t_n fires and p_{n+1} gets a token, then only t_1 is enabled again. Such an order can be repeated infinitely if p_{n+1} can always obtain tokens.

In the manufacturing environment, when designers want to specify a strict execution order among n possible choices, or specifications include a requirement, the above module can be used. For example, when two machines can process raw material, designers may start loading one machine first, and then the other. Such a process may be repetitive. However, a potential danger is that when the machine that is supposed to be loaded first cannot due to some unexpected reasons, then no other machines can start if the above decision-free choice PN design is adopted.

It is obvious that there is no real choice or conflict in such a module; thus, such a module is called a decision-free choice PN module.

5.3.2 Refinement Theory and its Extension

In order to present the refinement theory, replacement of a place or a transition in a Petri net by a subnet first needs to be defined.

Definition 5.5 *A replacement* of a place p' in $Z = (P \cup \{p'\}, T, I, O, m_0)$ with a place PN block $S_P = (\{p_{in}, p_{out}\} \cup P_s, T_s, I_s, O_s, m_{s0})$ results in a refined Petri net $Z' = (P', T', I', O', m_0')$ where

$$P' = \{p_{in}, p_{out}\} \cup P_s \cup P, T' = T_s \cup T,$$

$$m_0'(p) = \begin{cases} m_0(p) & \text{if } p \in P \\ m_{s0}(p) & \text{if } p \in \{p_{out}\} \cup P_s \\ m_0(p') + m(p_{in}) & \text{if } p = p_{in} \end{cases}$$

$$I'(p,t) = \begin{cases} I(p,t) & \text{if } p \in P \text{ and } t \in T \\ I_s(p,t) & \text{if } p \in P_s \text{ and } t \in T_s \\ 1 & \text{if } p = p_{out} \text{ and } I(p',t) = 1 \\ 0 & \text{otherwise} \end{cases}$$

$$O'(p,t) = \begin{cases} O(p,t) & \text{if } p \in P \text{ and } t \in T \\ O_s(p,t) & \text{if } p \in P_s \text{ and } t \in T_s \\ 1 & \text{if } p = p_{in} \text{ and } O(p',t) = 1 \\ 0 & \text{otherwise} \end{cases}$$

Similarly, a replacement of a transition t' in $Z = (P, T \cup \{t'\}, I, O, m_0)$ with a transition PN block $S_T = (Ps, \{t_{in}, t_{out}\} \cup T_s, I_s, O_s, m_{s0})$ results in a refined Petri net $Z' = (P', T', I', O', m_0')$ where

$$P' = P_s \cup P, T' = \{t_{in}, t_{out}\} \cup T_s \cup T$$

$$m_0'(p) = \begin{cases} m_0(p) & \text{if } p \in P \\ m_{s0}(p) & \text{if } p \in P_s \end{cases}$$

$$I'(p,t) = \begin{cases} I(p,t) & \text{if } p \in P \text{ and } t \in T \\ I_s(p,t) & \text{if } p \in P_s \text{ and } t \in T_s \\ 1 & \text{if } t = t_{in} \text{ and } I(p,t') = 1 \\ 0 & \text{otherwise} \end{cases}$$

$$O'(p,t) = \begin{cases} O(p,t) & \text{if } p \in P \text{ and } t \in T \\ O_s(p,t) & \text{if } p \in P_s \text{ and } t \in T_s \\ 1 & \text{if } t = t_{out} \text{ and } O(p,t') = 1 \\ 0 & \text{otherwise} \end{cases}$$

Thus, a place or transition is said to be *replaced* by a PN block.

Theorem 5.1 (Refinement) *Suppose that a place or a transition in Z is replaced by a basic design module and Z' is the refined net. Then,*

1. *Z' is bounded (safe) if and only if Z is bounded (safe);*

2. *Z' is live if and only if Z is live;*

3. *Z' is reversible if and only if Z is reversible.*

Here, a basic design module refers to a sequence PN module, a parallel PN module, a choice PN module, or a decision-free choice PN module.

Proof: To prove conclusions 1 and 2, one can directly use the results given by Valette [1979] or Suzuki and Murata [1983] by verifying that the above four basic design modules are well-formed or well-behaved. Using the method similar to ones in [Valette 79], the third conclusion can be showed.

5.4 Petri Net Modeling of Buffers

In the context of manufacturing, storage areas, stocks, or buffers are very common. For example, in a production line, a series of buffers between machines may exist. In a flexible manufacturing system, one may design an automatic storage and retrieval system for temporary storage of intermediate parts. Moreover, there are storage areas for raw materials and final products; temporary storage areas also exist for fixtures, pallets, or automatic guided vehicles. This section presents methods for modeling buffers in manufacturing systems using Petri nets. The resulting modules are useful in synthesizing Petri net models which maintain liveness, boundedness, and reversibility.

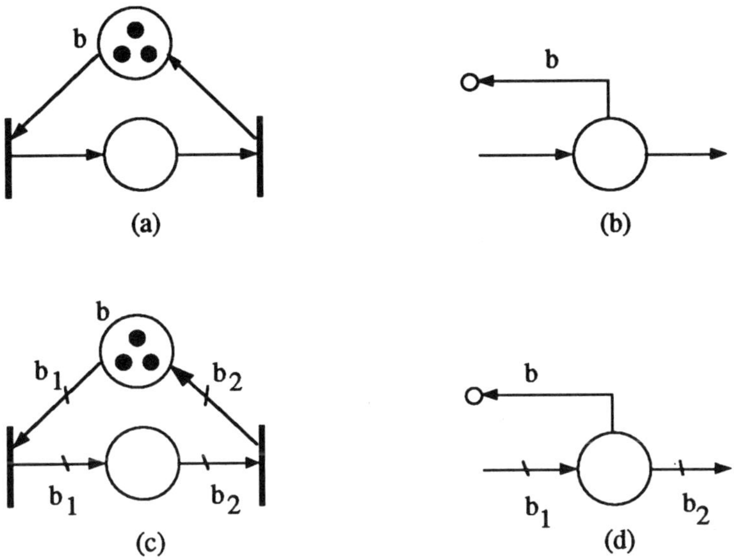

Figure 5.8: Two buffer modeling methods

5.4.1 Design of a Simple Buffer

In a manufacturing system, a buffer is a commonly used component and appropriate capacity of a buffer can lead to good performance of the system. Our objective is to model the buffer using a Petri net [Zhou 90e] rather than to find an optimal capacity of a buffer given a manufacturing system, e.g., a multiple machine and multiple buffer production line [Desrochers 89; Eleftheriu 89].

Conventional Methods

A simple buffer linked by two components, e.g., two machines, can be modeled as a buffer module using two places as shown in Fig. 5.8(a). Such modules are extensively used in the literature [Peterson 81; Zhou 89e].

When inhibitors are included in extended Petri nets, a buffer module is shown in Fig. 5.8(b). Such a module becomes a basis for a modular approach to analyzing the performance of production lines and networks [Al-Jaar 89, 90a].

Both modules have been extended to the case that each time b_1 slots or tokens are consumed upon firing of the left-side transition and b_2 slots become available upon firing of the right-side transition, as shown in Fig. 5.8(c) and (d). These modules capture the macro characteristics of a buffer using either two bounded places or a bounded place with an inhibitor arc. They assume that elements are randomly ordered. This implies that the elements can be viewed as the same type.

Safe Petri Net Modules for a Buffer

In order to model First-In-First-Out () buffers or to cope with special specifications (e.g., when a buffer holds two or more parts, some successive operations can be executed), *safe buffer Petri net modules* are introduced.

Agerwala [1979] provided a safe Petri net realization of a buffer of size two using four places and three transitions. The buffer is located between two stages of a pipeline. Such a case can be easily generalized to a buffer of capacity b, and the safe net module is constructed in Fig. 5.9(a) by using $2b$ places and $b+1$ transitions. It is noted that such a module assumes that elements are ordered and are subject to the First In First Out (FIFO) policy. In this realization, each part which enters the buffer and reaches its successive operation needs to fire b transitions, i.e., t_2 to t_{b+1}.

To avoid this, another Petri net module of a buffer with capacity b is designed and is depicted in Fig. 5.9(b) with initial marking $(1, 0, ..., 0)^T$ using $b+1$ places and $2(b+1)$ transitions.

If $b = 1$, then the buffer PN becomes a safe Petri net. In the following discussion, b is assumed to be greater than 1. For $i = 1, 2, ..., b+1$,

t_i: transition links to a precedent operation before the buffer;
r_i: transition links to a successive operation after the buffer;
p_i: place and if it is marked, the buffer holds $i-1$ parts.

Therefore, if p_1 is marked, then the buffer is empty. If p_{b+1} is marked, then the buffer is full. From the diagram, the following properties can readily be verified.

1. If p_1 is marked, then r_i are not enabled, or no successive operations can occur. Therefore, only a precedent operation can be executed.

2. If p_{b+1} is marked, then t_i are not enabled and only r_{b+1} is enabled. Therefore, only a successive operation can be executed.

3. For $i = 2, 3, ..., b+1$, t_i and r_{i-1} are enabled, but any other transition in this safe buffer Petri net module are not enabled.

4. Only one part in the buffer can be conveyed to the next operation for each firing of transition r_i and only one part can enter the buffer for each firing of transition t_i.

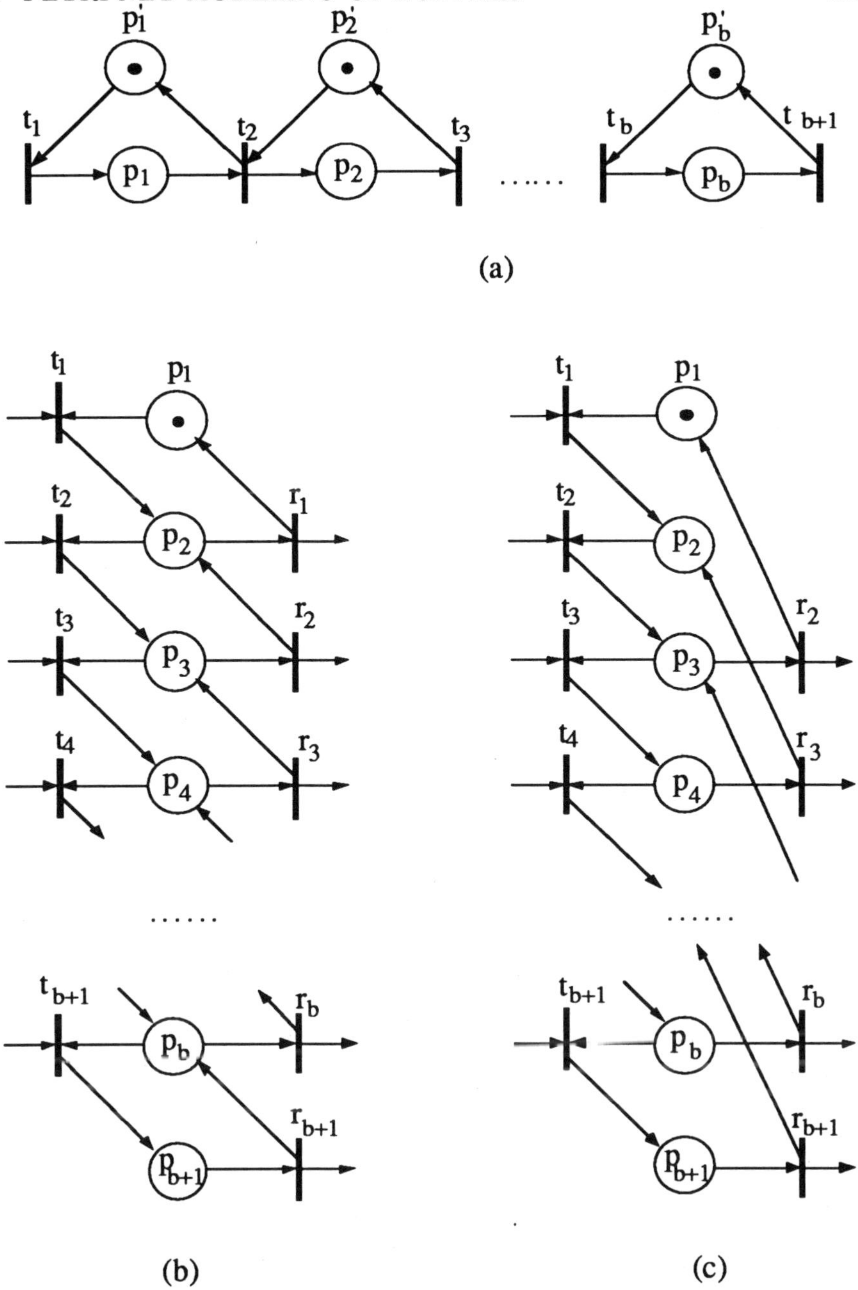

Figure 5.9: Realizations of a safe buffer module with capacity b

It is noted that elements are randomly ordered in this realization. Minor modification can result in the realizations which fit FIFO or LIFO (Last In First Out) if either inhibitors or complementary places are utilized.

Suppose that the workstation proceeding the buffer can produce a part each time while the workstation following the buffer needs to consume two parts each time; then, all the proposed safe buffer modules are no longer applicable. A realization is designed as shown in Fig. 5.9(c).

It is easily verified from the design that when a buffer holds a single part, i.e., $m(p_2) = 1$, no successive operation after the buffer can be performed. When the buffer holds two or more than two parts, a successive operation can be performed.

All these safe Petri net buffer modules offer a possibility to explore the internal operations of a buffer. They also provide another implementation scheme for buffers using only safe places.

Other work on modeling a buffer under the FIFO policy using colored Petri nets can be seen in [Martinez 86; Narahari 87].

5.4.2 Design of a Generalized Buffer

When a buffer is shared by several machines, its design can be depicted as shown in Fig. 5.10. The Petri net in Fig. 5.10(a) is used to model a shared buffer which can accommodate identical types of elements. Fig. 5.10(b) models a buffer which can accommodate different kinds of elements. Furthermore, the composition of Fig. 5.10(a) and (b) results in a Petri net module for a generalized buffer (see Fig. 5.10(c)) where k is the number of types of parts the buffer can hold.

5.4.3 Maintenance of System Properties

Theoretical results can be derived for the construction of a Petri net containing a buffer module. The addition of a bounded place and corresponding arcs, together with the occupying buffer places, e.g., $p_1, p_2, ..., p_k$ in Fig. 5.10(c), leads to a buffer module. Such an addition is proved to preserve boundedness, liveness, and reversibility under certain conditions.

The first two theorems hold true for any Petri net, whereas the third theorem is true only when a Petri net contains no sequential mutual exclusion structures; for example, there is no shared resource which can form an SME.

In the following theorems, ${}^{\bullet}p_i$ and $p_i{}^{\bullet}$ are *preset* and *postset* of the place p_i respectively. The preset of p is the set of all input transitions to the place p, i.e., ${}^{\bullet}p = \{t : t \in T \text{ and } O(p,t) = 1\}$. The postset of p is the set of all output transitions from the place p, i.e., $p^{\bullet} = \{t : t \in T \text{ and } I(p,t) = 1\}$.

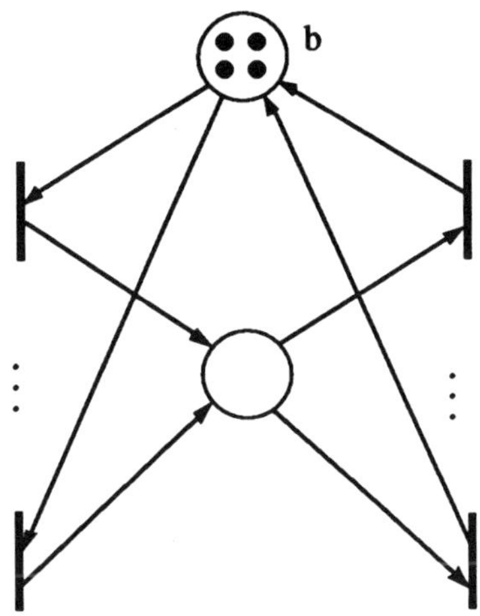

(a) Identical

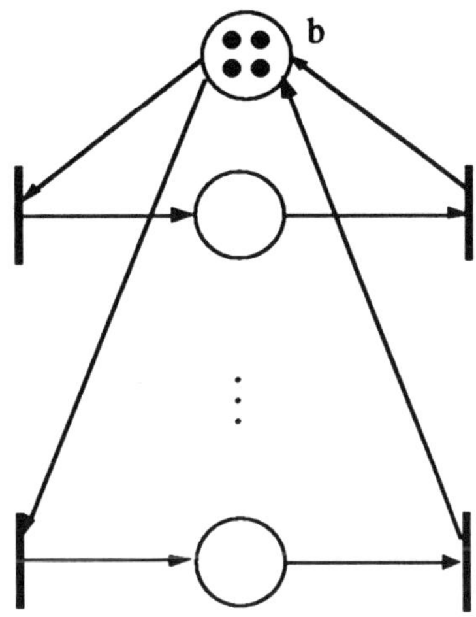

(b) Distinct types

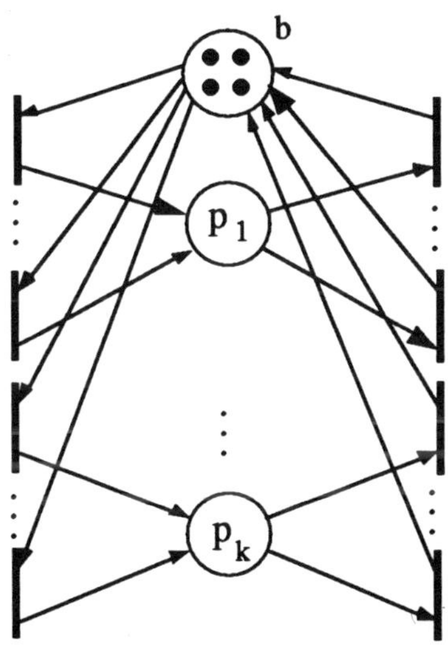

(c) Generalized

Figure 5.10: Shared buffer modules

Theorem 5.2 *Given* $Z = \{P \cup \{p_1, p_2, ..., p_k\}, T, I, O, m_0)\}$, $m_0(p_i) = 0, {}^{\bullet}p_i \neq \emptyset$, *and* $p_i{}^{\bullet} \neq \emptyset$ *for* $1 \leq i \leq k$. ${}^{\bullet}p_i \cap {}^{\bullet}p_j = \emptyset, p_i{}^{\bullet} \cap p_j{}^{\bullet} = \emptyset, i \neq j$, *for* $1 \leq i, j \leq k$. $Z' = (P', T', I', O', m_0'), P' = P \cup \{p_1, p_2, ..., p_k\} \cup p', T' = T$,

$$I'(p,t) = \begin{cases} I(p,t) & if\, p \neq p' \;\; and\, t \in T \\ 1 & if\, p = p' \;\; and\, t \in {}^{\bullet}p_1 \cup {}^{\bullet}p_2 \cup ... \cup {}^{\bullet}p_k \\ 0 & if\, p = p' \;\; and\, t \notin {}^{\bullet}p_1 \cup {}^{\bullet}p_2 \cup ... \cup {}^{\bullet}p_k \end{cases}$$

$$O'(p,t) = \begin{cases} O(p,t) & if\, p \neq p' \;\; and\, t \in T \\ 1 & if\, p = p' \;\; and\, t \in p_1{}^{\bullet} \cup p_2{}^{\bullet} \cup ... \cup p_k{}^{\bullet} \\ 0 & if\, p = p' \;\; and\, t \in p_1{}^{\bullet} \cup p_2{}^{\bullet} \cup ... \cup p_k{}^{\bullet} \end{cases}$$

$$m_0'(p) = \begin{cases} m_0(p) & if\, p \in P' - \{p'\} \\ b & if\, p = p' \end{cases}$$

Then p' *and* p_i *for* $1 \leq i \leq k$ *are b-bounded.*

Proof: Since transitions in ${}^{\bullet}p_i$ can fire at most b times without firing transitions in $p_i{}^{\bullet}$. Thus p_i can have at most b tokens at any marking for $1 \leq i \leq k$. QED

Theorem 5.3 *Suppose that two marked Petri nets Z and Z' are constructed as in Theorem 5.2. If (a)* $\forall m \in R(Z, m_0), m(p_1) + m(p_2) + \cdots + m(p_k) \leq b$, *or (b)* $\forall t \in p_1{}^{\bullet} \cup p_2{}^{\bullet} \cup \cdots \cup p_k{}^{\bullet}, |{}^{\bullet}t| = 1$, *then*

1. *Z' is bounded (safe) if Z is bounded (safe);*
2. *Z' is live if Z is live; and*
3. *Z' is reversible if Z is reversible.*

Proof: For (a), the addition of such a resource place that is initially marked with b tokens does not affect any firing sequences of the transitions. For (b), after firing any transitions in ${}^{\bullet}p_i$ will lead to the marking which enables every transition in $p_i{}^{\bullet}$ which will be enabled unless one of them fires. Also, firing any transition in $p_i{}^{\bullet}$ will mark p' which enables those transitions in ${}^{\bullet}p_i$ in Z. Thus any firing sequences in Z can be constructed in Z' to achieve a desired marking. Therefore, Conclusions 1 to 3 follow. QED

Theorem 5.3 reveals the fact that if a buffer is never full or if each transition in the post-set of places $p_1, p_2, \ldots, p_k$ has one input place only, then the system exhibits the same properties as the original one. The theorem is very easy to apply. However, when the condition becomes no longer true, this theorem is not applicable. Therefore, Theorem 5.4 is developed provided that no sequential mutual exclusion structure exists in the concerned buffer modules in Fig. 5.10(a) although the theorem is not generally true for the buffer modules shown in Fig. 5.10(b) and (c).

Definition 5.6 *A sequential mutual exclusion (SME) structure* $S = (P_s, T_s, I_s, O_s, m_{0s})$ is defined as a subnet of $Z = (P_A \cup P_B \cup P_C, T, I, O, m_0)$. Z and S satisfy the following condition:

1. $|P_s| \geq 1$, $|T_s| \geq 0$, and if $|P_s| > 1$ or $|T_s| > 0$, $\exists$ a place $x \in P_s$ such that $\forall y \in P_s \cup T_s - \{x\}$, $EP(y,x) \cup EP(x,y) \neq \emptyset$;

2. $\exists x \in P_s$, and four different transitions $t, t', r, r' \in T - T_s$, $\ni EP(x,t) \neq \emptyset$, $EP(t',x) \neq \emptyset$, $EP(x,r) \neq \emptyset$, $EP(r',x) \neq \emptyset$, $EP(t,r') \neq \emptyset$, and t', $r \in EP(t,r')$.

We say that Z contains a sequential mutual exclusion structure. For example, a single place p_6 constitutes such an SME structure since there are t_1, t_2, t_3, t_4 which satisfies Condition 2 by letting $x = p_6$ in this example.

Theorem 5.4 *Suppose that two marked Petri nets Z and Z' are constructed as in Theorem 5.2 with $k = 1$. If Z contains no SME structure as defined in Definition 5.6, then*

1. *Z' is bounded (safe) if Z is bounded (safe);*

2. *Z' is live if Z is live; and*

3. *Z' is reversible if Z is reversible.*

Proof: Using the notation adopted in the proof of Theorem 3.1, we have: $m(p')$ is assumed to be the last element of any marking $m \in R(Z', m_0')$. To save space,

$$m = \begin{pmatrix} \mu \\ m(p') \end{pmatrix}$$

is written as $(\mu, m(p'))$. In addition,

$$Q(Z', m_0') = \{m : (m, m(p')) \in R(Z', m_0')\}.$$

Since each sequence of transitions f fires in Z', it can be applied to Z. Therefore, $Q(Z', m_0') \subset R(Z, m_0)$. Thus since p' is b-bounded according to Theorem 5.2, Z' is bounded if Z is so. The following proof of liveness is conducted by contradiction. Similarly, the proof of reversibility can be constructed by contradiction. If Z satisfies the condition in Theorem 5.3, then we

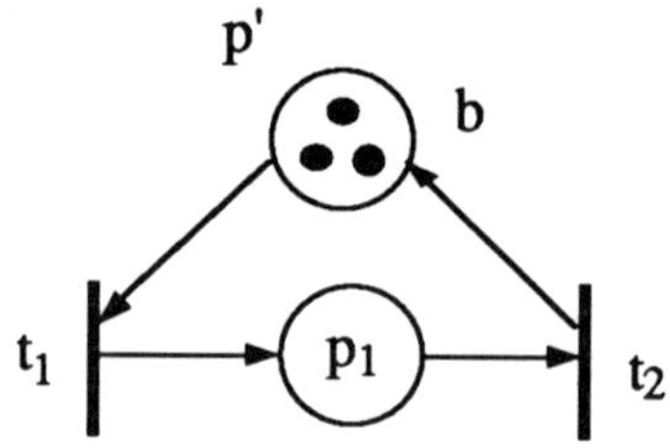

Figure 5.11: A simple buffer of capacity b in the proof of Theorem 5.4

do not need to prove this theorem. Therefore, the condition in Theorem 5.3 is assumed to be not true. For simplicity, consider the case shown in Fig. 5.8, i.e., $|{}^\bullet p_1| = |p_1{}^\bullet| = 1$. Let $\{t_1\} = {}^\bullet p_1$, $\{t_2\} = p_1{}^\bullet$ as shown in Fig. 5.11.

Suppose that Z is live and Z' is not live. This implies that there is at least a transition which is live in Z, but not live in Z'. We derive an SME structure in Z. The difference between Z and Z' implies that if there is a transition that is not enabled from $m_0'[f> m$, then $m(p') = 0$. From m, neither t_1 nor t_2 is enabled. For t_2, $\exists p$, its marking will be zero no matter which sequence of enabled transitions from m fires and $EP(p, t_2) \neq \emptyset$. Since Z is live at m, this means that $\exists$ at least one transition t' such that $EP(t', t_1) \neq \emptyset$ and its firing marks a place p'' and $EP(p'', p) \neq \emptyset$. Certainly there are two parallel paths from t to t_2: $EP(t', t_1) \neq \emptyset$ due to $t_1 p_1 t_2$, and $EP(t', p'')$, $EP(p'', p)$, and $EP(p, t_2)$. If there are no other transitions which are connected to $EP(p'', p)$ then one can easily verify that Z' is also live. Thus by tracing elementary paths back from t' and forward from t_2, one can find a pair of transitions say t and r' such that $EP(t, t') \neq \emptyset$, $EP(t_2, r') \neq \emptyset$, $EP(p, t) \neq \emptyset$ and $EP(r', p'') \neq \emptyset$.

Let $x = p$. The above p, t, t', t_2, and r' constitute an SME structure in Z according to Definition 5.6. This is a contradiction.

The similar proof can be constructed for the cases shown in Fig. 5.10(a). QED

Theorem 5.4 is not generally true for cases presented in Fig. 5.10(b) and (c). One can show counter-examples. Fortunately, the synthesis procedure provided next gives a mechanism to produce a net with a property required by one of these three theorems, provided that the first-level Petri net contains no SME structures, e.g., the two basic kinds of Petri nets as discussed in Section 5.2.

5.5 Synthesis Procedure Description

The proposed method treats operation places and resource places separately, such that one can construct a Petri net using top-down methods in the initial stage, and using bottom-up methods in the final stage. Two major benefits are that

1. The details can be introduced in an incremental way, such that complexity can be alleviated; and

2. The important properties of resulting Petri nets are guaranteed, making costly analysis for some qualitative properties unnecessary.

This section describes a hybrid synthesis methodology using Petri nets. Two stages are identified, top-down design and bottom-up design. Top-down design aims at refining the net step by step so as to include enough system operation details for the purpose of implementation. Bottom-up design focuses on a correct construction of interactions among subsystems or detailed operation processes. All shared resources are considered at this stage and become a key to constructing a Petri net with desirable qualitative properties. This method is illustrated by Petri net modeling of an automated manufacturing system in Chapter 6 and a flexible manufacturing system in Chapter 7.

5.5.1 Top-down Design

Top-down design of Petri nets refers to a design process for Petri nets using stepwise refinement of places, transitions, or substructures. This is accomplished by replacing them with more complicated subnets. Each successive step contains increasing detail until the implementation level is reached. Refinement theory is used to guarantee that each step of refinement preserves the desirable system properties of Petri nets. Since those refinements are often local and required to have no side-effects, interaction among subnets is very difficult to handle using such a strategy. Therefore, bottom-up design is also adopted and is discussed in the next section.

The first step to initiate top-down design is to find an appropriate Petri net as a first-level model for the system. This model needs to catch crucial relationships such as the choice-synchronization structure. This is because such kinds of relationship cannot easily be constructed in the later design process, no matter what methods are used. At this point, overall analysis of a system to be modeled and its specification are often necessary. The decomposition of a large system into independent subsystems is also completed at this stage.

Depending on the specification, a variable resource places set is determined and the resources represented in this set are often necessary to start any operation in a system if they exist. For example, raw materials, fixtures, and pallets are necessary to start a production system.

Next, operations involved in a subsystem are modeled by sequence, parallel, choice, and decision-free choice Petri net modules. In this decomposition process, an operation may be divided into several sub-operations until further division is unnecessary. It is obvious that each refinement increases the number of operation places, i.e., A-places, which model operation processes.

Next, the structure of a unique operation place plus transitions is replaced by a structure with resource places, operation places, and transitions. This corresponds to the addition of fixed resource places. At this stage, only non-shared resources, i.e., B-places, are considered.

The refinement theory discussed in Section 5.3.2 is used to guarantee that the replacement of a place or a transition with a basic design module preserves the properties: boundedness, liveness, and reversibility. At the same time, the theorems developed in Chapters 3 about mutual exclusion are used to maintain the properties of Petri nets if each added B-place forms a 1-PME. Such a verification is very easy to obtain.

The above discussions are summarized as the following steps, which constitute the top-down design stage:

Step 1. Choose a bounded, live, and reversible Petri net as a first-level model of a system that is working when all major resources are available, and determine P_A, P_B, and P_C where P_A and P_B are subject to increasing while P_C is often fixed and its members have a variable number of initial tokens.

Step 2. Decompose this system into several subsystems expressed as operation places using the defined basic design modules. Replace these operation places by more detailed basic design models until no operations can be further divided or until one reaches a point at which the further division is not necessary. P_A is increased at each step.

Step 3. Appropriately add the non-shared resource places at each stage when one or several operations require the resources. The arcs are linked to the resource places so that the conditions in Corollary 3.1 are satisfied. At each addition only P_B is augmented.

It is noted that the refinement of operations and the addition of non-shared resource places can be alternately applied. This means that no strict order needs to be followed between them.

5.5.2 Bottom-up Design

As we mentioned in the last section, top-down design is difficult in resolving complicated interaction problems. These problems are often critical to design a deadlock-free system. This section applies the bottom-up design idea to the construction of Petri nets. Specifically, mutual exclusion theory developed is used to resolve resource-sharing modeling problems. Since every net for each sub-system already exists after the top-down design stage, what is left is to design correct interactions among the existing sub-nets or among detailed operation processes in a net. From the design point of view, such a process is called bottom-up design .

First, consider shared buffers if they exist using the theorems established in Section 5.4.3, the appropriate construction of buffer modules and existing Petri nets ensures that Petri nets are bounded, live, and reversible.

Next, design fixed shared resources which form generalized parallel mutual exclusions one by one to satisfy the given specification. These resources are shared by independent or concurrent processes. Often, the difficulty of verifying whether a shared resource is appropriately added increases as a Petri net contains more generalized parallel mutual exclusions (GPMEs).

Finally, consider each resource which is shared by sequentially related groups of processes. If it makes a generalized sequential mutual exclusion (GSPM), then its token capacity and related parameters are computed by deriving and using the defined subnets of the whole Petri net. Feasible initial markings are determined by the theorems developed in this book so that the final Petri net has desirable properties.

During the above process, any shared resource's failure to form a GPME or GSME often leads to a finding of either an inappropriate specification or an incorrect design of the net. Such early findings can be used to adjust the specifications or to give an insight into some built-in deadlocks contained in the system. They also prevent effort to correct some design errors which may be detected in a global net through existing analysis methods. The reasons why a GPME or GSME cannot be formed for a shared resource in a system can often offer some interesting approaches to avoiding possible deadlocks. This was illustrated in modeling a two-robot assembly line in Chapter 3.

Following the steps given in the last section, the design process of this stage can be stated as follows:

Step 4. Add each buffer place which, coupled with the existing A-place to represent a buffer, forms a Petri net buffer module if any buffers exist. Each added buffer place with initial tokens whose number equal the capacity of the buffer becomes a B-place.

Step 5. Add the shared resource places which form PMEs or GPMEs. The resource places become B-places.

Step 6. Add the shared resource places which form an SME or GSME and calculate the achievable token capacities between their PMEs or GPMEs, α, β, and γ; and determine the number of initial tokens in p, $\forall p \in P_C$ by using the theorems developed in Chapter 4. The resource places are added to P_B.

Finally, we get a bounded, live, and reversible Petri net model with a set of feasible initial markings.

Remark 1. One can add arcs to shared resource places stepwise in Step 4 and Step 5.

Remark 2. One can also conduct stepwise refinement using basic design modules at each step. Verified modules can also be used in in the refinement process in additional to the proposed modules.

Remark 3. It is interesting to note two facts here if only the first three steps described in the top-down stage are applied. One is that a bounded, live, and reversible marked graph can be easily synthesized, beginning with any choice-free net presented in Section 5.2.1. The other is that there are two ways to synthesize a non-persistent Petri net. One is to begin with a non-persistent bounded, live, and reversible Petri net, for example, a choice-synchronization net as discussed before; another is to refine a place or transition with a choice or decision-free choice PN design module in Step 2. Therefore, the synthesis procedure characterized by our first design stage extends the work of marked graphs [Murata 80a, b; Jonsonbaugh 81] to ordinary Petri nets.

Remark 4. The complete procedure which includes top-down and bottom-up design stages extends the existing synthesis approaches [Suzuki 83; Valette 79; Zhou 89b] to more general cases by incorporating resource-sharing cases.

Chapter 6

Illustration of Hybrid Synthesis

6.1 An Automated Manufacturing System

This automated manufacturing system consists of the following major components: two entries, two exits, five machines, two buffers, four robots, and a two AGV system as shown in Figure 6.1.

Entries: There are two entries for two types of raw materials, F and G, which are produced into two different kinds of products. Each raw material piece is fixtured to a pallet so that it can be transferred using robots and the AGV system.

Exits: There are two exits for finished F-parts and G-parts, respectively. The output process is accomplished via two AGVs.

Machines: Machines M_1 and M_2 process F-raw pieces from Entry 1. Both machines produce identical F-parts which are sent to Buffer 1. These intermediate parts are further machined by M_3 to produce finished F-parts. M_4 processes G-raw pieces from Entry 2 and produces G-parts which are stored in Buffer 2. M_5 machines those intermediate G-parts from Buffer 2.

Buffers: Buffer 1 of size b_1 receives F-parts from M_1 or M_2. It outputs these intermediate F-parts one by one to M_3. Buffer 2 of size b_2 receives processed G-parts from M_4. It outputs these G-parts one by one to M_5.

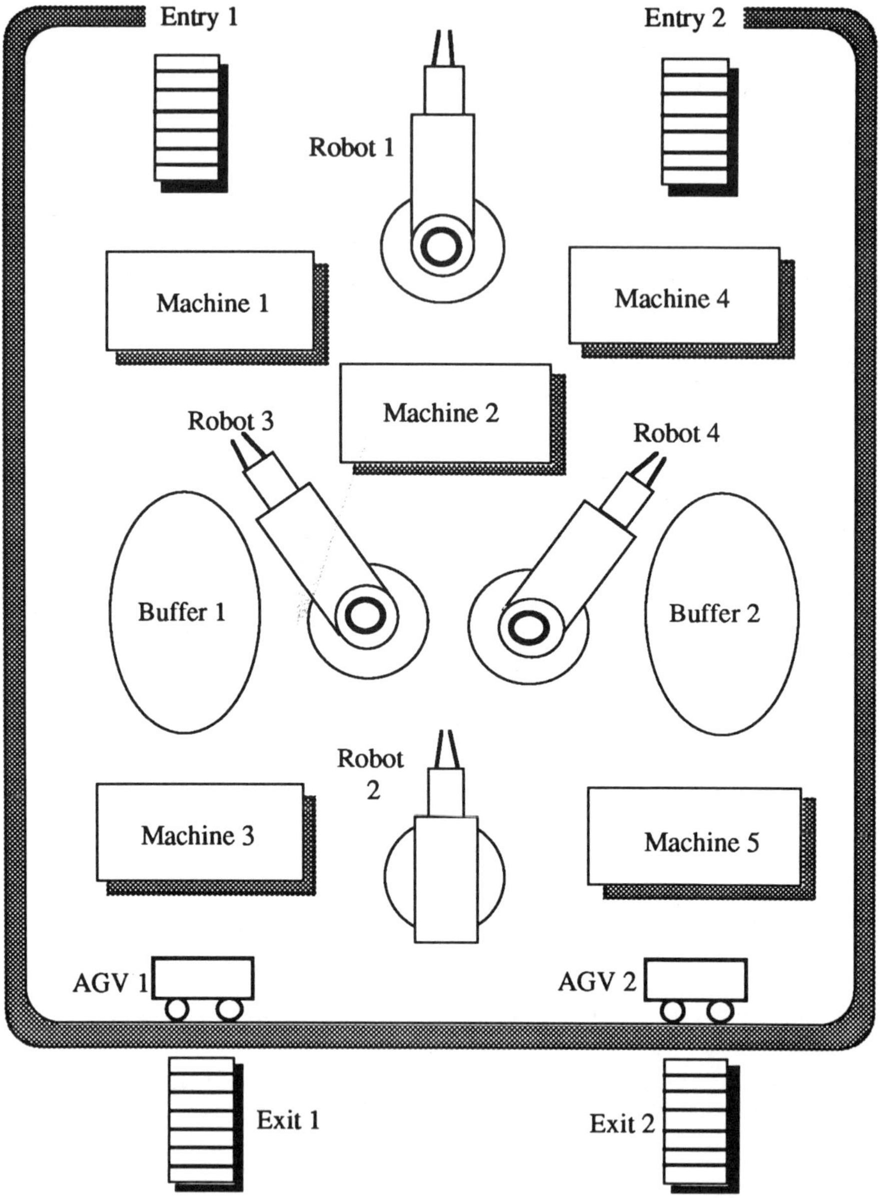

Figure 6.1: Layout of an automated manufacturing system

Robots: Robot R_1 loads M_1, M_2, and M_3, and none of these three machines has any priority to use this robot. In a non-deterministic manner, R_1 chooses one machine if raw material is available and the machine is ready.

Robot R_2 unloads M_3 and M_5, and sends finished parts fixtured to pallets to the AGV system.

Robot R_3 is shared by M_1, M_2, and M_3. It performs the following functions: unloading M_1 and M_2, conveying type-F intermediate parts to and from Buffer 1, and loading M_3. Any F-parts processed by M_1 or M_2 must be stored in Buffer 1 before any further processing. Such a design prevents any just-processed parts by M_1 or M_2 from intermediate processing by M_3.

Similarly, Robot R_4 is used to serve M_4 and M_5. It can be used to unload M_4, to deliver type-G intermediate parts to and from Buffer 2, and to load M_5. Buffer 2 plays the same role as Buffer 1 does.

AGV System: Two AGVs are designed for the delivery of final parts and the release of pallets in the system. From M_3, AGV1 sends final F-parts to the output and pallets back to Entry 1; from M_5, AGV2 sends final G-parts to the output and pallets back to Entry 2. Since they take different paths, collision is avoided (see in Fig. 6.1). Both AGVs can also work concurrently after Robot 2 puts parts on them one by one.

An unlimited source of raw materials is assumed. Once machines, robots, or AGVs start work on any operation, they cannot be interrupted until the work is complete. The requirement is that the system is not deadlocked and has the smallest amount of starving by designing a deadlock-free Petri net-based control structure. Therefore, we must construct a Petri net model for the system and determine the maximum number of initial pallets which are allowed in the net while guaranteeing the net's boundedness, liveness and reversibility.

6.2 Modeling Process

The design process of a Petri net model for the system follows the procedure proposed in Section 5.5. There are two stages: top-down design and bottom-up design which include the six steps as follows:

1. Design of a first-level Petri net

2. Decomposition and refinement of operations

3. Addition of non-shared resources

4. Addition of buffer places

5. Addition of shared resources to form GPMEs and

6. Addition of shared resources to form GSMEs.

6.2.1 Design of a First-level Petri Net

Without consideration of shared resources, i.e., all robots, this system can be viewed as two independent subsystems called the left subsystem and the right subsystem. The left subsystem consists of three machines, a buffer, and an AGV. The right subsystem consists of two sequentially linked machines, a buffer, and an AGV. Since no choice-synchronization structures are involved in these two subsystems, a first-level Petri net is designed as a model having two separate Petri nets as shown in Fig. 6.2 . The two nets are choice-free nets, as discussed in Section 5.2.1. Initially, we have:

$$P_A = \{p_{01}, p_{02}\};$$
$$P_B = \emptyset;$$
$$P_C = \{p_{10}, p_{20}\}.$$

The variable resource places, i.e., C-places p_{10} and p_{20} represent the availability of pallets with F- and G-raw materials, respectively. Operation places p_{01} and p_{02} represent the working of the left and right subsystems, respectively. Transitions always represent the start or the end of some operation in all cases. The net is bounded, live, and reversible whenever $m_0(p_{10}) > 0$, and $m_0(p_{20}) > 0$.

It is noted that any place with its subscript starting with a zero is to be refined. For example, p_{01} and p_{02} are refined in Fig. 6.3 whose discussion follows next.

6.2.2 Decomposition and Refinement of Operations

In order to fully describe the system, further decomposition and refinement are required. Since either subsystem can be decomposed into three stages to generate a final product, both p_{01} and p_{02} in the first-level net model in Fig. 6.2 are replaced with a three-place sequence module, respectively. The refined net is obtained in Fig. 6.4, where

$$P_A = \{p_{01a}, p_{1b}, p_{01c}, p_{02a}, p_{2b}, p_{02c}\}.$$

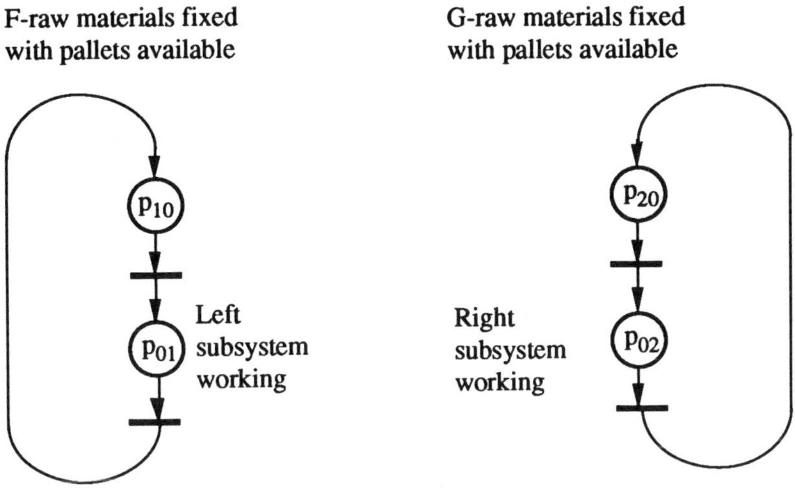

Figure 6.2: A first-level Petri net model consisting of two separate nets

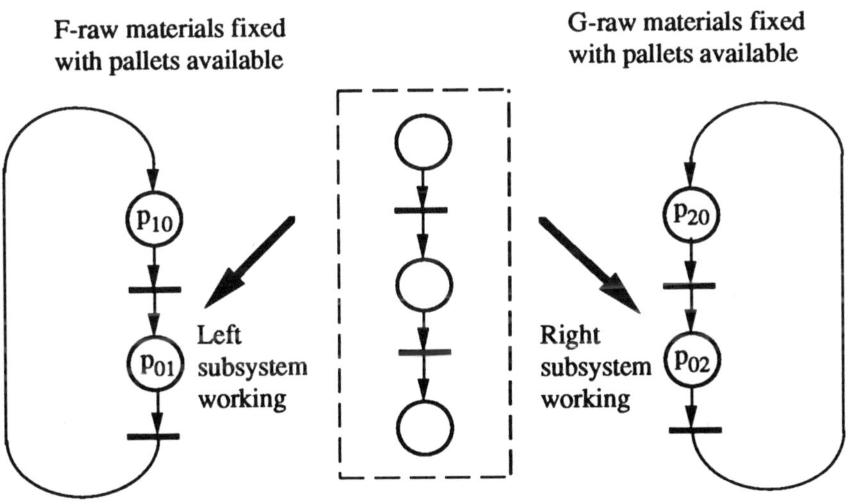

Figure 6.3: Two places p_{01} and p_{02} are to be refined

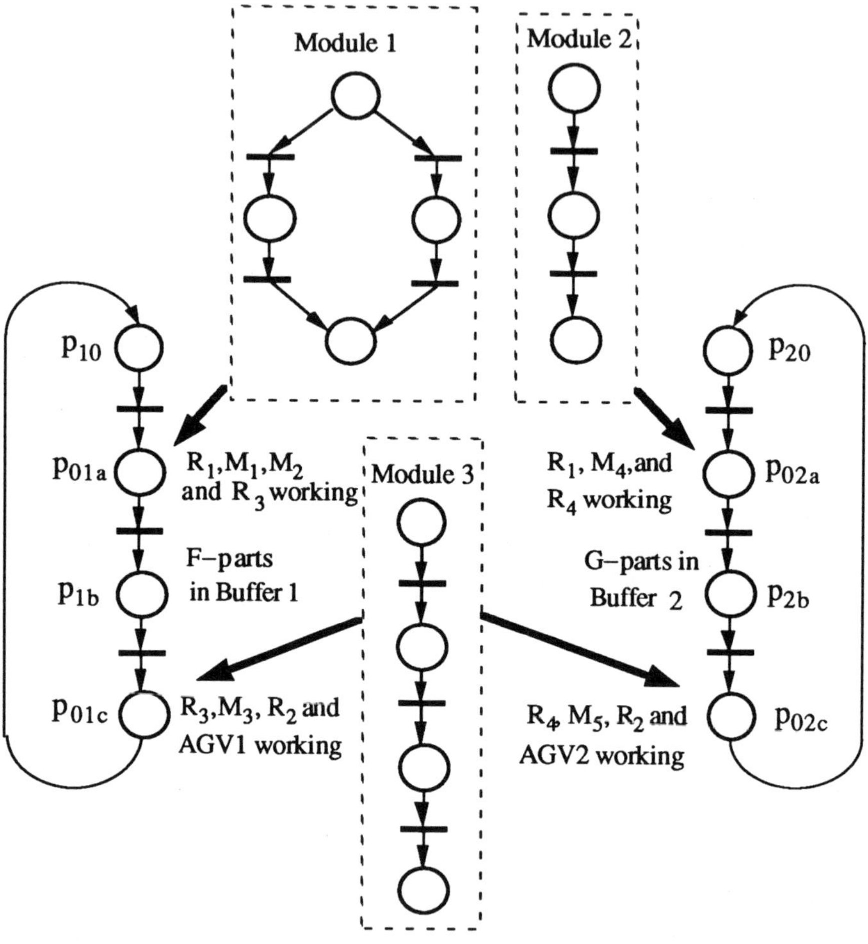

Figure 6.4: The Petri net after refinements of places p_{01} and p_{02}

Table 6.1: Place labels for the Petri net in Fig. 6.4

Place	*Interpretation*
p_{10}	F-type raw materials, each fixed with a pallet, are available
p_{01a}	R_1, M_1, M_2 and R_3 working
p_{1b}	F-parts in Buffer 1
p_{01c}	R_3, M_3, R_2 and AGV1 working
p_{20}	G-type raw materials, each fixed with a pallet, are available
p_{02a}	R_1, M_4 and R_4 working
p_{1b}	F-parts in Buffer 1
p_{02c}	R_4, M_5, R_2 and AGV2 working

The meaning of each place is explained shown in Fig. 6.4 and Table 6.1. Further decomposition and refinement are still needed:

1. Place p_{01a} is replaced with a choice PN module (Module 1 in Fig. 6.4);
2. Place p_{02a} is replaced with a three-place sequence PN module (Module 2 in Fig. 6.4);
3. Both p_{01c} and p_{02c} are replaced with a four-place sequence PN module (Module 3 in Fig. 6.4);
4. Places p_{01b} and p_{02b} model intermediate F- and G-parts which are stored in Buffer 1 and Buffer 2, respectively. No further refinement for these two places is needed.

The Petri net which is refined through operation places is depicted as shown in Fig. 6.5. The operation place set is

$$P_A = \{p_{ij}, p_{19}, i = 1, 2, \text{ and } 1 \leq j \leq 8\}.$$

The place interpretations are given in Table 6.2.

6.2.3 Addition of Non-shared Resources

The five machines and two AGVs are used by a single operation process and, therefore, are non-shared resources in the system. All the additions of these resource places are shown in Fig. 6.6. The newly added places belong to a fixed resource place set.

$$P_B = \{p'_{12}, p'_{13}, p'_{17}, p'_{19}, p'_{22}, p'_{26}, p'_{28}\}.$$

These B-places, each of which is initially marked with one token, are explained as in Table 6.3:

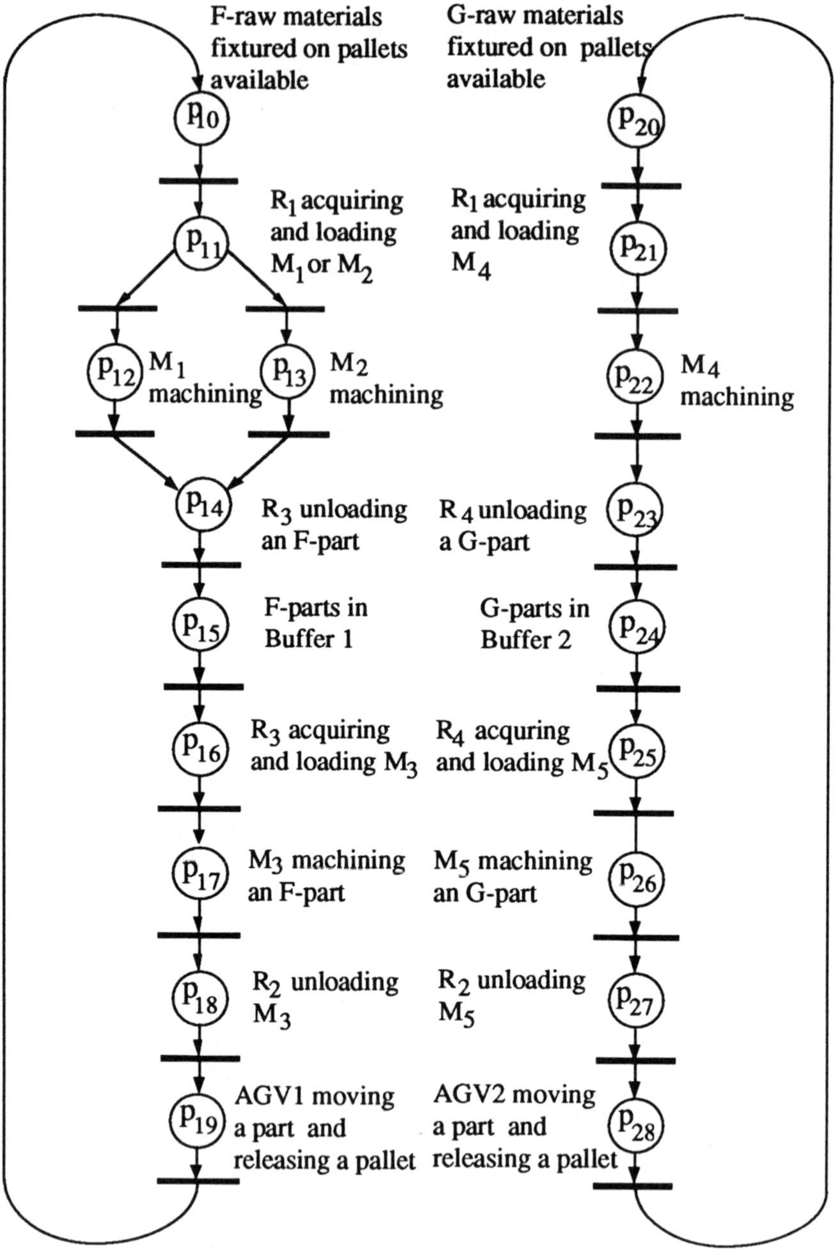

Figure 6.5: The Petri net after refinements of $p_{01a}, p_{02a}, p_{01c}$, and p_{02c}

Table 6.2: Place labels for the Petri net in Fig. 6.5

Place	*Interpretation*
p_{10}	F-type raw materials, each fixtured on a pallet, are available
p_{11}	R_1 acquiring a pallet from Entry 1 and loading M_1 or M_2
p_{12}	M_1 machining an F-raw piece
p_{13}	M_2 machining an F-raw piece
p_{14}	R_3 unloading M_1 or M_2 and placing an intermediate F-part in Buffer 1
p_{15}	Buffer 1 storing intermediate F-parts
p_{16}	R_3 acquiring a pallet from Buffer 1 and loading M_3
p_{17}	M_3 machining an intermediate F-part
p_{18}	R_2 unloading M_3 and putting a pallet on AGV1
p_{19}	AGV1 moving a finished F-part to Exit 1 and releasing a pallet to Entry 1
p_{20}	G-type raw materials, each fixtured with a pallet, are available
p_{21}	R_1 acquiring a pallet from Entry 2 and loading M_4
p_{22}	M_4 machining a G-raw piece
p_{23}	R_4 unloading M_4 and placing an intermediate G-part in Buffer 1
p_{24}	Buffer 1 storing intermediate G-parts
p_{25}	R_3 acquiring a pallet from Buffer 2 and loading R_5
p_{26}	M_5 machining an intermediate F-part
p_{27}	R_2 unloading M_5 and putting a pallet on AGV2
p_{28}	AGV2 moving a finished G-part to Exit 2 and releasing a pallet to Entry 2

Table 6.3: Place labels for the Petri net in Fig. 6.6

Place	*Interpretation*
p'_{12}	M_1 is ready
p'_{13}	M_2 is ready
p'_{17}	M_3 is ready
p'_{19}	AGV1 is ready
p'_{22}	M_4 is ready
p'_{26}	M_5 is ready
p'_{28}	AGV2 is ready

The refinement theory and the non-shared resource addition theorem guarantee the properties of the net in Fig. 6.6. These properties include boundedness, liveness, and reversibility whenever there are positive numbers of tokens in both variable resource places, i.e., p_{10} and p_{20}. Tokens in these two C-places represent pallets with raw materials to be processed.

6.2.4 Addition of Buffer Places

This system consists of two simple buffers. Either is designed as a simple buffer PN module in the Petri net model as shown in Fig. 4.7. The theorems developed in Section 5.4.3 guarantee the preservation of system properties achieved by the Petri net in Fig. 6.6. Two B-places p'_{15} and p'_{24} have the explanation:

p'_{15}: Empty slots in Buffer 1 are available if there are tokens in it; and
p'_{24}: Empty slots in Buffer 2 are available if there are tokens in it.

In particular, since p'_{15}, p_{15} and their two related transitions satisfy the second condition in Theorem 5.3, the construction of this resulting buffer module preserves the properties of the net, i.e., boundedness, liveness, and reversibility. The same argument is applicable to p'_{24}, p_{24} and their two related transitions. It is clear that

$$m_0(p'_{15}) = b_1$$

and

$$m_0(p'_{16}) = b_2$$

6.2.5 Addition of Shared Resources to Form GPMEs

First, consider the robot R_2 shared by two machines, i.e., M_3 and M_5. After p_2 and related arcs are added as shown in Fig. 6.8, it is obvious that $(p_2, (t_9, t_{10}), (t_{27}, t_{28}))$ is a 2-PME. Therefore, the net is bounded, live, and reversible.

Second, consider the robot R_1 shared by M_1, M_2, and M_4. Since M_1 and M_2 share a common entry, after R_1 acquires an F-raw material, it can load either M_1 or M_2 when both are idle. This shared resource is modeled in Fig. 6.8 using a B-place p_1. It is also easy to verify that $(p_1, \{(t_1, \{t_2, t_3\}), (t_{21}, t_{22})\})$ forms a GPME. Therefore, the theory ensures boundedness, liveness, and reversibility of the net in Fig. 6.8 whenever there are tokens in both C-places.

Two B-places p_1 and p_2, each marked initially with one token, have the following explanation:

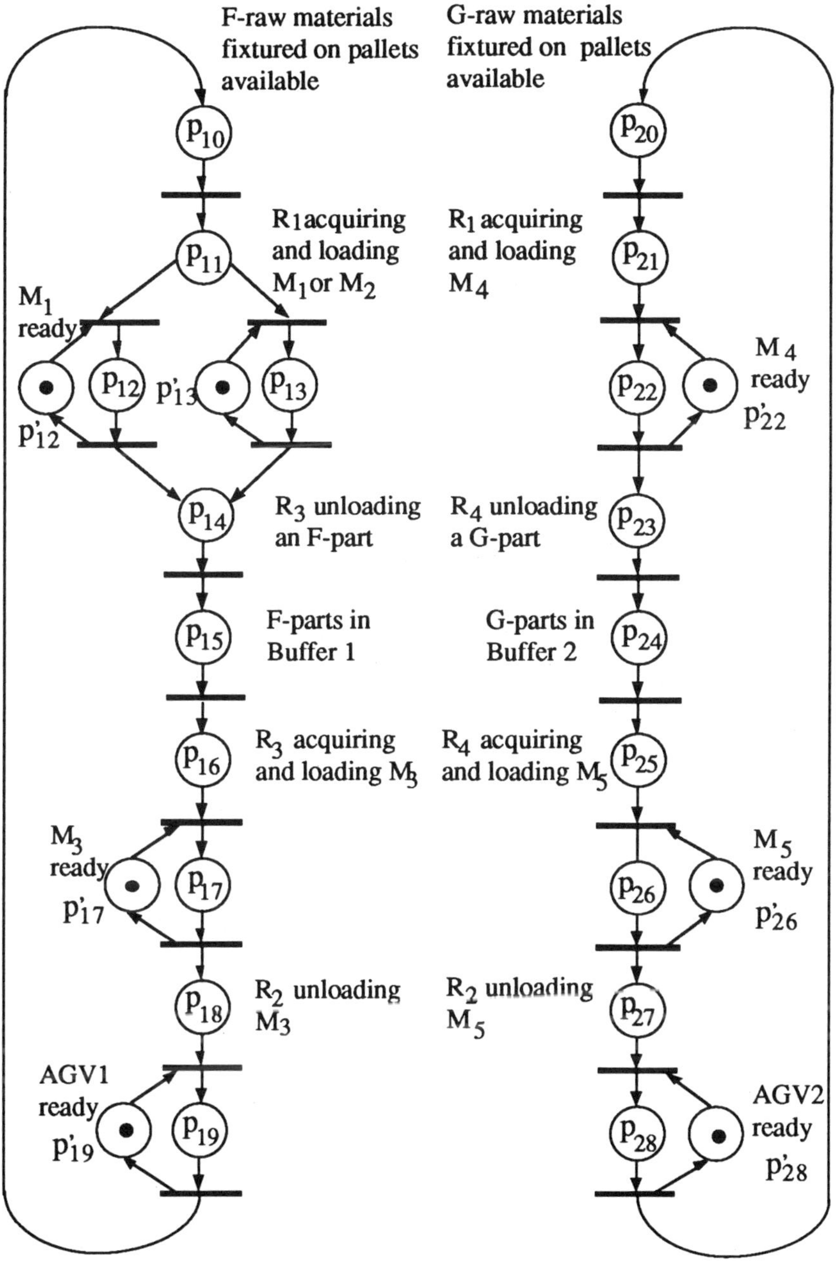

Figure 6.6: The Petri net after addition of non-shared resources

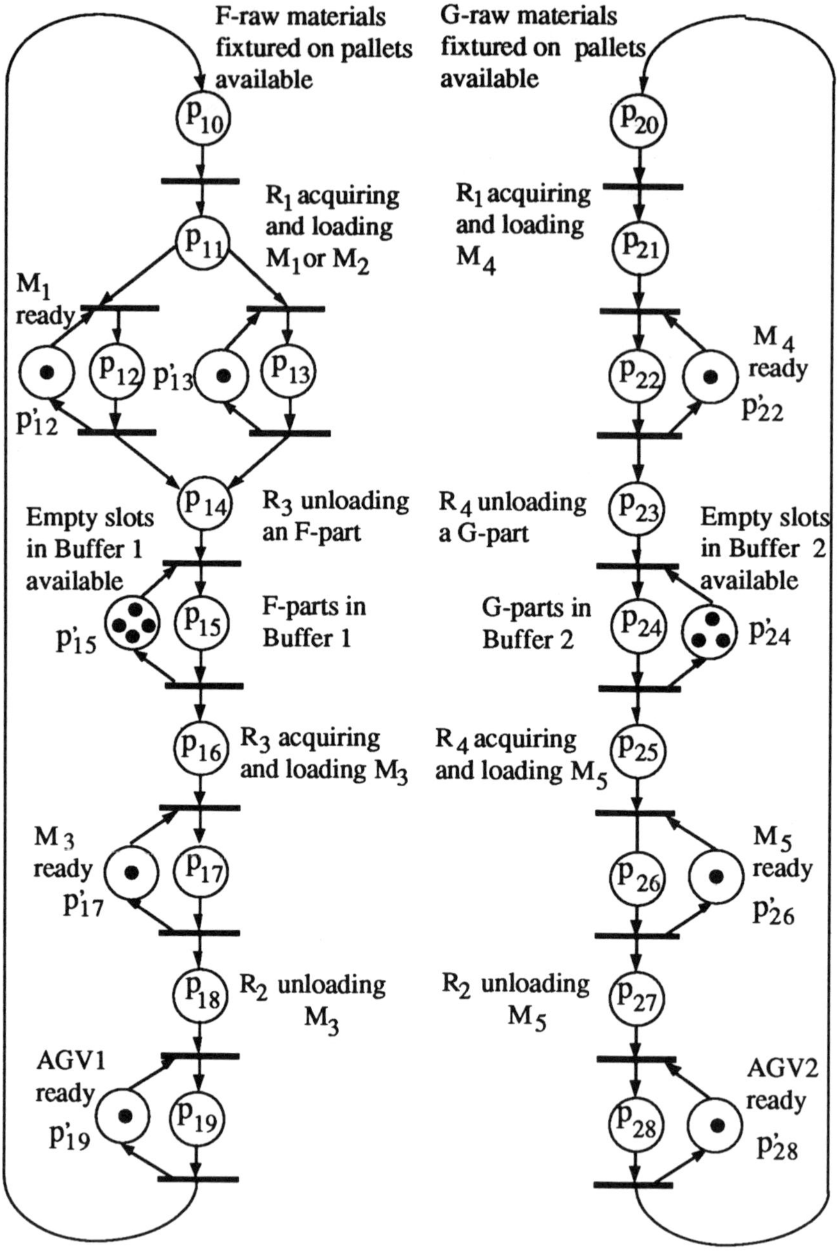

Figure 6.7: The Petri net after addition of buffer resource places

p_1: R_1 is ready; and
p_2: R_2 is ready.

6.2.6 Addition of Shared Resources to Form GSMEs

The most difficult part is to deal with sequentially shared resources. This system provides two such resources, i.e., Robots R_3 and R_4. For example, R_4 first needs to unload M_4, and then it may either unload M_4 again or load M_5. *Unloading-M_4* and *loading-M_5* are two sequentially-related processes. Although the final version of the Petri net model for the system is shown in Fig. 6.9, the modeling processes can be divided into several steps.

First, consider the shared robot R_3. When only arcs $(p_3, t_4), (p_3, t_5)$, and (t_6, p_3) are added to the net, a 1-GPME is formed, i.e., $(p_3, \{(\{t_4, t_5\}, t_6)\})$. When arcs (p_3, t_7) and $(t_8, p3)$ are added, the structure $(p_3, \{(\{t_4, t_5\}, t_6)\} \cup \{(t_7, t_8)\})$ can be verified readily to be a GSME for this system. The results of its token capacity, α, β, and γ are as follows:

$$C^1 = C(t_6, t_7) = b_1.$$

$$\alpha(p_{10}, t_4) = \alpha(p_{10}, t_5) = 1, \alpha(p_{20}, t_4) = \alpha(p_{20}, t_5) = 0;$$
$$\alpha(p_{10}) = 1, \alpha(p_{20}) = 0.$$
$$\beta(p_{10}, t_4) = \beta(p_{10}, t_5) = 1, \beta(p_{20}, t_4) = \beta(p_{20}, t_5) = 0;$$
$$\beta(p_{10}) = 1, \beta(p_{20}) = 0;$$
$$\gamma(p_{10}, t_4) = \gamma(p_{10}, t_5) = m_0(p_{10}).$$

Therefore,

$$\gamma(t_4) = \gamma(t_5) = \infty.$$

Using Theorems 4.7 for this GSME, we have the feasible initial markings:

$$1 \leq m_0(p_{10}) \leq b_1,$$

and

$$1 \leq m_0(p_{20})$$

which are required for liveness and reversibility.

Second, Robot R_4 is considered. The structure $(p_4, \{(t_{23}, t_{24})\} \cup \{(t_{25}, t_{26})\})$ can be easily verified to be an SME. Its token capacity, α, β,andγ are readily computed as follows:

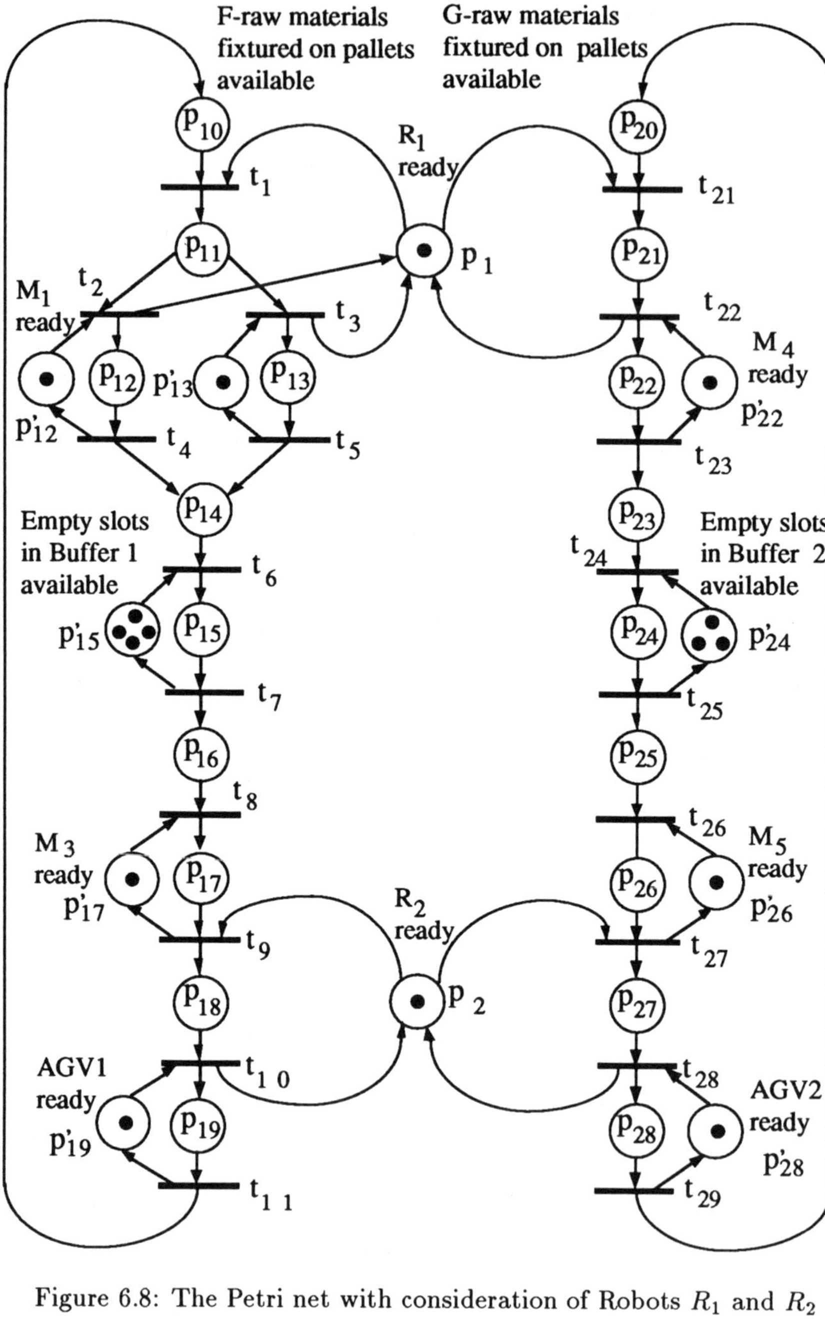

Figure 6.8: The Petri net with consideration of Robots R_1 and R_2

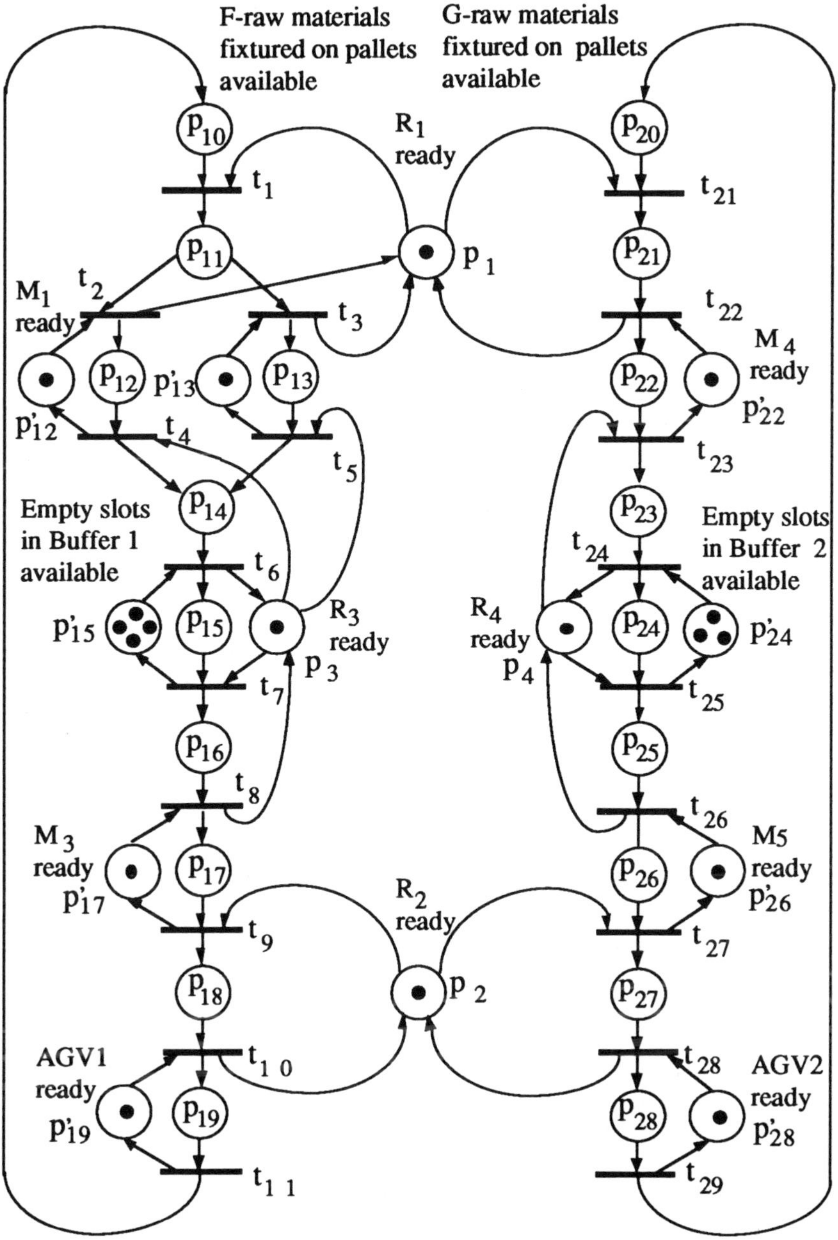

Figure 6.9: The final Petri net model for the system

$$C^1 = C(t_{24}, t_{25}) = b_2.$$

$$\alpha(p_{10}) = 0, \alpha(p_{20}) = 1.$$

$$\beta(p_{10}) = 0, \beta(p_{20}) = 1;$$

and

$$\gamma(p_{20}, t_{23}) = m_0(p_{20}).$$

Therefore,

$$\gamma(t_{23}) = \infty$$

Using Theorems 4.2-4 for this SME, we obtain:

$$1 \leq m_0(p_{20}) \leq b_2$$

and

$$1 \leq m_0(p_{10})$$

Therefore, in order to guarantee the desirable qualitative properties of the net as shown in Fig. 6.9, the initial markings have to satisfy:

$$1 \leq m_0(p_{10}) \leq b_1$$

and

$$1 \leq m_0(p_{20}) \leq b_2$$

With these feasible initial markings and the Petri net structure shown in Fig. 6.9, the Petri net is bounded, live, and reversible. The operation place set, fixed resource place set, and variable resource place set can be summarized as follows:

$$P_A = \{p_{1i}, 1 \leq i \leq 9\} \cup \{p_{2i}, 1 \leq i \leq 8\}$$

$$P_B = \{p_1, p_2, p_3, p_4, p_{12'}, p_{13'}, p_{15'}, p_{17'}, p_{19'}, p_{22'}, p_{24'}, p_{26'}, p_{28'}\}$$

and

$$P_C = \{p_{10}, p_{20}\}.$$

In order to decrease possible starvation in an operational environment, we choose:

$$m_0(p_{10}) = b_1$$

and

$$m_0(p_{20}) = b_2.$$

6.3 Discussion

Using the proposed hybrid method, one can synthesize, step by step, a bounded, live, and reversible Petri net model for a manufacturing system. The determination of appropriate initial token distribution is also possible so that the net has desirable properties given its structure.

If needed, the refinement of a place or transition can be conducted at any stage. For example, place p_{19} in the final net in Fig. 6.9 may be replaced with a two-place sequence PN module which uses two sequentially-related processes to model the function of p_{19}. The desirable properties are still preserved. Other examples by applying the hybrid synthesis methodology can be found in [Zhou 90a, b, 92a, b] and the FMS examples related to this synthesis method in [Chao 92; Jeng 92; Zhou 92c; Zurawski 92].

We can prevent the systems from being deadlocked by limiting the number of pallets or fixtures in the system. However, for the above system, if the speeds of the machines are very different, then starving may occur. For example, when M_5 is much faster that M_4, then M_5 will starve from time to time. Under such a situation, one may need to add one more machine which can do the same job as M_4. Then the Petri net model needs to be modified. Such modification may destroy the desirable properties. Further research is needed to study the conditions that can preserve the properties of the Petri net model. Some augmentation methods will be introduced and discussed in Chapter 8 in the context of manufacturing system error recovery. When we have several types of jobs which need processing by different machines in a flexible manufacturing system, all machines will become shared resources with some sharing of robots and material movement systems among different machines. For such a system, the synthesis of a Petri net with desirable properties will become more and more difficult. In other words, the incorporation of planning or scheduling into the current approaches will become a new thrust in the synthesis of Petri nets.

Chapter 7

Discrete Event Control of FMS

7.1 Introduction

This chapter presents the synthesis and implementation of a Petri net (PN) model used for the discrete-event control of a moderately sized flexible manufacturing system (FMS). This system includes machining, assembly, materials movement, and storage, and comprises two machining stations, one assembly station, three robots, four conveyors, two dedicated storage devices, and an automated storage and retrieval system. The chapter is written to demonstrate that the research in PNs over the last three decades has matured such that meaningful industrial applications are now becoming possible.

The objectives of this chapter are

1. To describe the FMS physical simulator laboratory for which the PN is synthesized and a supervisory controller constructed and operated.

2. To show that the hybrid methodology in Chapter 5 for the synthesis of bounded, live, and reversible PN models is robust enough to handle the complexity of a medium sized FMS; and

3. To illustrate a straightforward method for constructing a supervisor, based on the net, which can be directly implemented and executed.

The practical significance of this research is that to the authors knowledge, it represents the most complex synthesized PN that guarantees qualitative properties and that has also been used for the control of a physical system.

7.2 The Flexible Manufacturing System

The flexible manufacturing system (FMS) developed at Rensselaer Polytechnic Institute and shown in Fig. 7.1 is an FMS physical simulator which is a 1/6 scale model of an automated factory shop floor and can manufacture a family of products [Weiss 86; Aslam 88]. It was originally developed to illustrate the benefits of constructing a scaled physical simulation before committing to a full scale production facility. It is also used for research on the design and implementation of control for automated manufacturing systems using PNs [Rudolph 89; Zhou 90c, 92b].

The system takes raw input stock of two types, machines them into desired shapes, and then assembles these two finished parts into a product. One part starts as a rectangular solid block and is milled and drilled on the CNC milling machine and ends as a geometrically shaped part with a cylindrical hole in its top. The other part starts as a cylindrical solid peg and is turned by the CNC lathe into a complex cylindrical/conical shaped peg whose base fits into the hole of the finished block part. The robotic insertion of the peg into the hole of the block is the assembly process. The family of products for this FMS comprises various sizes and shapes of blocks and pegs assembled in this manner.

7.2.1 Layout of the FMS

The major components of the system as illustrated in Fig. 7.1 are:

1. The CNC mill and drill machine.

2. The CNC lathe.

3. The Microbot robot is a shared resource used to load and unload the materials between the CNC lathe and Conveyor 3 (C3), and between the CNC milling machine and C2.

4. The Automated Storage and Retrieval System (ASRS) is a buffer where pallets with raw materials and intermediate parts can be stored. The ASRS structure is composed of 5×4 storage locations including 19 usable pallet-storage bins. An elevator carries loaded pallets to and from the ASRS.

5. The Automated Material Transfer System (AMTS) is a set of shared resources which include four two-way conveyors (one pallet at a time) with presence sensors at each end and a Gantry robot. The AMTS is in charge of the transfer of the materials between the different stations. Universal pallets are used , i.e., they can hold all materials, parts and

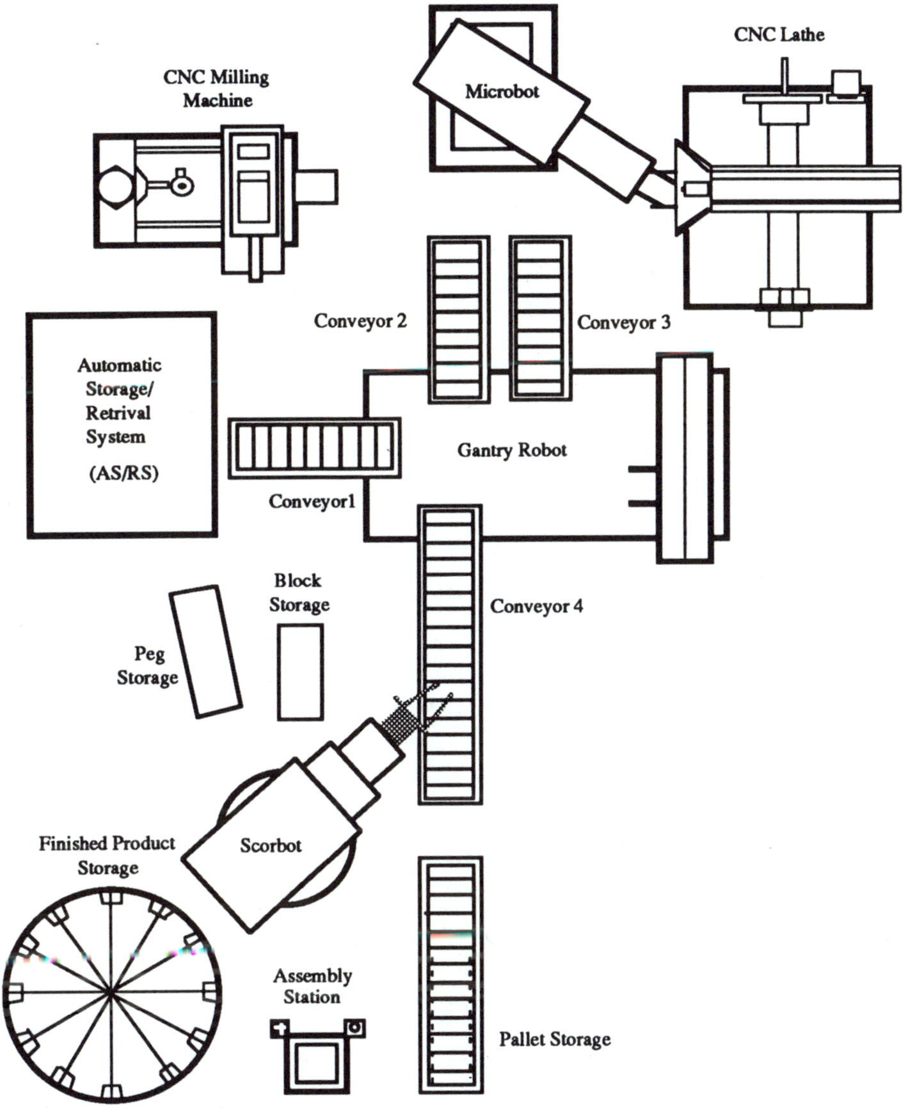

Figure 7.1: The layout of the FMS simulator

products. Gantry robot: transfers pallets, holding either raw stock or a finished part, between the four conveyors.

C1: transfers pallets holding either raw stock or a finished part between the ASRS and the Gantry robot.

C2: transfers pallets holding either raw stock or a finished part between the CNC milling machine and the Gantry robot.

C3: transfers pallets holding either raw stock or a finished part between the CNC lathe and the Gantry robot.

C4: transfers pallets holding either raw stock from the raw material storage area to the Gantry robot or finished parts from the Gantry robot to the assembly area.

6. The Scorbot robot is a shared resource with several functions: moves pallets between the empty pallet storage and C4 in the AMTS; moves pegs from the peg storage and blocks from the block storage onto pallets on C4; moves blocks and pegs from pallets on C4 to the assembly cell and assembles them; and moves the assembled product to the carousel (finished product storage).

7.2.2 FMS Computer Control Structure

Figure 7.2 illustrates the implemented computer control structure of the FMS simulator. At the top level is the system supervisor, an IBM PC-XT. On the middle level are 6 APPLE II computers which serve as workstation controllers. The lowest level consists of the FMS components described above, which are controlled by the computers at the middle level. By executing a PN program, the supervisory controller coordinates operations between the various components and initiates operations by communicating with the appropriate mid-level workstation computers. One can see that the present system uses a centralized control architecture. The system is modifiable to investigate other distributed control structures.

7.2.3 Material Flow in the FMS

The following is a typical flow of the activities necessary to transform raw materials into a finished product. The parentheses indicate an alternate set of objects for the same activity. Some of these activities may occur concurrently or in a different order [Zhou 92b].

1. The Scorbot robot moves an empty pallet from the gravity fed storage to C4.

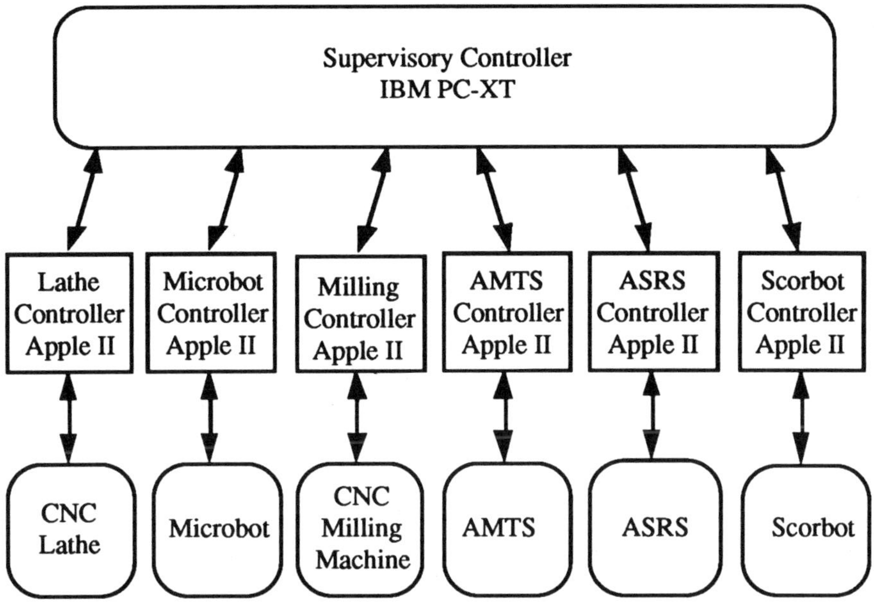

Figure 7.2: A computer control architecture

2. The Scorbot robot takes a raw block (peg) from the block (peg) storage and places it in the empty pallet on C4.

3. C4 moves the loaded pallet to the Gantry robot.

4. The Gantry robot moves the pallet from C4 to C2 (C3).

5. C2 (C3) moves the pallet to the CNC milling machine (lathe).

6. The Microbot robot takes the raw block (peg) from the pallet and loads it into the CNC milling machine (lathe).

7. The CNC milling machine (lathe) fixtures the raw block (peg) and machines the part.

8. The Microbot robot unloads the finished block (cylindrical) part from the CNC milling machine (lathe) to the pallet on C2 (C3).

9. C2 (C3) moves the pallet to Gantry robot.

10. Gantry robot moves the pallet from C2 (C3) to C4.

11. C4 moves the pallet with the finished block (cylindrical) part to the Scorbot robot.

12. The Scorbot robot takes the finished block (cylindrical) part and places it in the assembly cell. If it is a cylindrical part it assembles the final product.

13. The Scorbot robot moves the finished product to the output carousel.

14. The Scorbot robot moves the empty pallet from C4 to the pallet storage.

While this is the typical flow to produce one output, the reader is reminded that the system will have concurrent work in process for a number of product outputs. The PN based supervisor discussed later will coordinate and synchronize all activities and resources for the FMS.

7.3 Design of a Petri Net for the FMS

7.3.1 Design of a Macro-level Petri Net Model

To obtain a first-level Petri net for this system, one needs to analyze the major activities in terms of choices and precedence relations. For the given FMS, recall that the assembly process is the insertion of a turned peg part into the hole of a machined block part. For assembly to occur, a block part must be loaded into the assembly station before the peg part. This precedence can be enforced by the type of structure as we discussed in Chapter 5. In this FMS this precedence relation can be traced back to the beginning of the process when the system must choose whether to first load a pallet with raw material for a peg part or a block part. A simplified macro-level net is shown in Fig. 7.3 and Table 7.1, which is structurally the same as the Petri net discussed in Section 5.2.2 and shown in Fig. 5.2(d). This example shows a system with raw materials and pallets, p_1, two processes, p_2 and p_3, each producing different parts which are then assembled in a specified order, t_4 and t_3.

After two different parts are produced, modeled by t_1-p_2 and t_2-p_3, respectively, they are assembled to produce a final product. Place pairs p_4, p_5 and p_6, p_7 form part of the choice-synchronization structure guaranteeing that a block will be produced first, followed by a peg and that they will enter the assembly station in the same order. This net is bounded, live, and reversible when $m_0(p_1) \geq 2$ and the initial token distribution over the other places is as shown in Fig. 7.3.

Based on the simple Petri net shown in Fig. 7.3 and an analysis of the major system activities, we can design a macro-level net as shown in Fig. 7.4 by

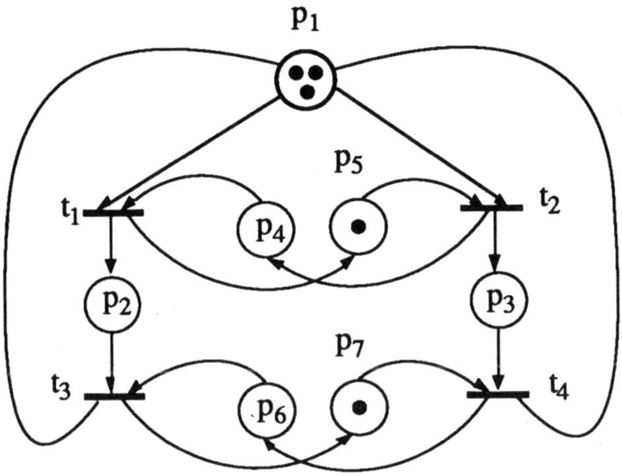

Figure 7.3: The Petri net model which contains a choice-synchronization structure where $m_0(p_1) \geq 0, m_0(p_2) = m_0(p_3) = m_0(p_4) = m_0(p_6) = 0$, and $m_0(p_5) = m_0(p_7) = 1$

Table 7.1: Place and transition labels for Petri net shown in Fig. 7.3

p_1: Pallets with raw material available
p_2: Processing a peg
p_3: Processing a block
p_4: Peg processed second
p_5: Block processed first
p_6: Block available at the assembly cell
p_7: Assembly cell available
t_1: Start processing a peg
t_2: Start processing a block
t_3: Send a block to the assembly cell
t_4: Assemble a block and a peg

taking into account the shared ASRS. Table 7.2 provides the interpretation for the places and transitions for this macro-level PN. Considering the following expansion process for the left side of Fig. 7.3. First, t_1 in Fig. 7.3 is replaced by a choice structure t_1-p_4-$t_3(t_4)$-p_6-t_7 in Fig 6, as framed by a box with dotted lines, and t_3 is similarly replaced by a choice structure. The same expansions apply to the right side of the net. Places p_2 and p_3 in Fig. 7.3 become p_8 and p_9 in Fig. 7.4. Next, we evolve the control pair of places, p_4 and p_5, in Fig. 7.3 into p_{16} and p_{17} in Fig. 7.4. These guarantee the alternative delivery of the raw materials, blocks and pegs. The other control pair, p_{14} and p_{15}, which evolve from p_6 and p_7 in Fig. 7.3, guarantee that a block reaches the assembly station first. They also guarantee that only one block and one peg are allowed to reach the assembly cell for each final product. In order to model the availability of raw pegs and blocks, we have created two C-places, i.e., p_2 and p_3, in Fig. 7.4. In this net the operation place set $P_A = \{p_i, 4 \leq i \leq 16, i \neq 15\}$, the fixed resource place set $P_B = \{p_{15}, p_{17}\}$, and the variable resource place set $P_C = \{p_1, p_2, p_3\}$. This net can be easily shown to be live, reversible, and bounded when $m_0(p_1) \geq 2, m_0(p_2) \geq 1$, and $m_0(p_3) \geq 1$. Further refinements of the places and transitions in this net are discussed and illustrated in the next section. It is noted that transitions t_4, t_5, t_{12}, and t_{13} represent the ASRS activities.

7.3.2 Stepwise Refinement of the Petri Net

Using top-down stepwise refinement of places and transitions, the Petri net in Fig. 7.4 can lead to a Petri net with sufficient operational detail. The final net as shown in Fig. 7.6 at the end of this section is obtained by adding all resource places as described in the next section.

Consider the two following examples which illustrate refinement of a place and a transition. The first one is the refinement of an operation place p_8 in Fig. 7.4, i.e., machining a peg on the lathe. It can be replaced by a sequence module which consists of loading the lathe, processing a peg; and unloading the lathe. In Fig. 7.6, the module is represented by the places and transitions: $P_{36}, T_{13}, P_{43}, T_{14}, P_{47}$, as framed by a box with round corners. We use capital P and T instead of p and t in Fig. 7.6 to preserve legibility.

Transition t_4 in Fig. 7.4, for example, can be replaced by a sequence transition block consisting of eight transitions T_{16} to T_{23} and seven places, P_{21}, P_{25} , P_{29}, P_{33}, P_{38} , P_{44}, and P_{48}, as framed by a box. Similarly, we refine the t_5, t_{12}, t_{13} in Fig. 7.4 to get the corresponding refined net shown in the other three middle boxes in Fig. 7.6. The net in Fig. 7.4 is further expanded by refining other transitions and places. P_{60}, P_{77} and P_9, P_{10} correspond to control pairs p_{14}, p_{15} and p_{16}, p_{17} in Fig. 7.4.

Note that all places and transitions in Fig. 7.6 are renumbered and different

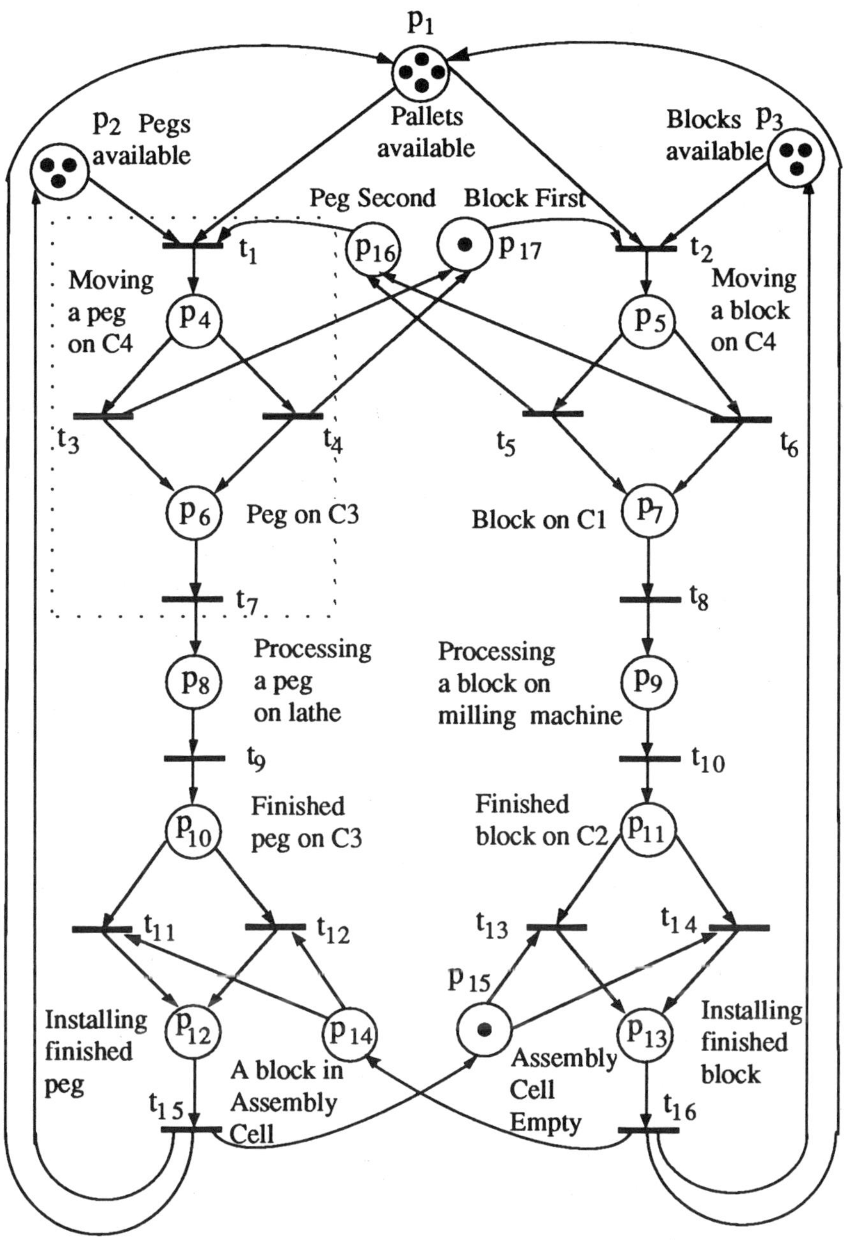

Figure 7.4: A macro-level Petri net model

Table 7.2: Place and transition labels for the macro-level Petri net

p_1: Pallets available
p_2: Pegs available
p_3: Blocks available
p_4: Peg-carrying pallet on C4
p_5: Block-carrying pallet on C4
p_6: Peg-carrying pallet on C3
p_7: Block-carrying pallet on C1
p_8: Machine raw peg on lathe
p_9: Machine raw block on milling machine
p_{10}: Finished-peg-carrying pallet on C3
p_{11}: Finished-block-carrying pallet on C2
p_{12}: Install finished peg at assembly station
p_{13}: Install finished block at assembly station
p_{14}: Block in a assembly cell available
p_{15}: Assembly cell available
p_{16}: Control place, a peg enters the system following a block
p_{17}: Control place, a block enters the system preceding a peg
t_1-t_2, t_7-t_{10}, t_{15}-t_{10}: Start or completion
t_3 (t_6): Move a peg (block) to C3 (C1)
t_4 (t_5): Move a peg (block) to C3 (C1) through ASRS
t_{10} (t_{14}): Move a finished peg (block) to C3 (C1)
t_{12} (t_{13}): Move a finished peg (block) to C3 (C1) through ASRS

from those in Fig. 7.4.

7.3.3 Modeling of Resources

Dedicated resources such as the lathe and milling machine are simply added to the model as illustrated by fixed resource places such as P_{37} and P_{41}. The following discussion focuses on the shared resources including the ASRS, the Scorbot robot, the Microbot robot, and the AMTS which is composed of the following independent shared resources: the Gantry robot, C1, C2, C3, and C4.

The ASRS is a shared buffer which can accommodate four kinds of palleted parts, i.e., raw-peg, raw-block, finished-peg, and finished-block. A generalized buffer can be modeled as shown in the simple example in Fig. 7.5, which is a particular case of the net as shown in Fig. 5.10(b). Similarly, we obtain the generalized buffer Petri net module for the ASRS as shown in Fig. 7.6. The places and transitions are: P_{28}; P_{21}, P_{25}, P_{29}, P_{33}; P_{22}, P_{26}, P_{30}, P_{34}; P_{54}, P_{58}, P_{63}, P_{67}; P_{55}, P_{59}, P_{64}, P_{68}; and T_{16-20}; T_{24-28}; T_{47-50}; T_{55-59}. This is done in two steps: refinement of transitions t_4, t_5, t_{12}, and t_{13} as discussed in the last subsection and addition of P_{28} and the related arcs with its initial

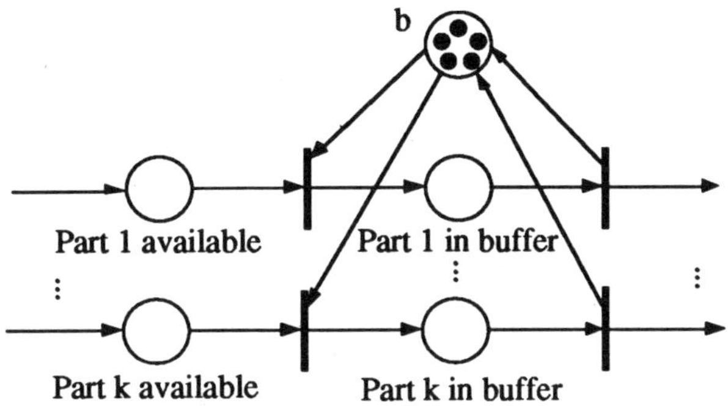

Figure 7.5: A simple generalized buffer module

marking $m_0(P_{28}) = 19$ to represent ASRS capacity. This construction of the buffer module satisfies Condition (b) in Theorem 5.3. Therefore, the boundedness, liveness, and reversibility of the net remain unchanged.

The Microbot robot (P_{31}) is shared by the lathe and the milling machine in parallel. It is also sequentially shared by the loading and unloading processes for each of them. Thus P_{31} and its related transitions form a sequential mutual exclusion in the net.

We have also modeled the other sequential mutual exclusions for the Scorbot robot, the Gantry robot, C1, C2, C3, and C4. For example, the Scorbot robot and its related transitions constitute a sequential mutual exclusion. Its functions include the transfer of all raw materials to the system for processing and from the system for final assembly. We compute the token capacity, 21, between these two group of processes. This includes the buffer size, 19, plus the capacities of the lathe and milling machine, 2, for this system. The other parameters are as follows:

$$\alpha(P_1) = 1,\ \alpha(P_6) = 1,\ \alpha(P_7) = 1$$

$$\beta(P_1) = 0.5,\ \beta(P_6) = 1,\ \beta(P_7) = 1$$

Then, using sequential mutual exclusion theory in Chapter 4, i.e., Theorems 4.6 and 4.7, and noting that $m_0(P_1) \geq 2$ is a necessary condition for liveness and reversibility of the intermediate nets, we obtain:

$$2 \leq m_0(p_{pallet}) \leq 21,$$

$$1 \leq m_0(p_{peg}),$$

and

$$1 \leq m_0(p_{block}).$$

where $P_1 = p_{pallet}$, $P_6 = p_{peg}$, and $P_7 = p_{block}$, and p_{pallet} is easily verified to be a principal variable resource place. The reason it is not necessary to limit the initial number of blocks and pegs is that the total number of pallets determines how many parts at most can enter the system. It is also noted that given $m_0(p_{pallet}) = 21$, in order to decrease the possible starving of machines, we should choose:

$$m_0(p_{peg}) = 10$$

and

$$m_0(p_{block}) = 10.$$

The result for the Petri net developed here shows that when $m_0(p_{pallet}) \geq 22$, $m_0(p_{block}) \geq 11$, and $m_0(p_{peg}) \geq 11$, the system will have a deadlock. This has a very good intuitive explanation: when the above condition exists, the system can have the following situation: the buffer has 19 positions full, 10 for blocks, 9 for pegs, the mill is processing a block, the lathe is processing a peg, and the Scorbot robot loads an empty pallet (to carry a raw-peg) into the system at C4. Then the system will deadlock. The condition imposed on the initial marking of this net excludes such an undesirable state. The hybrid methodology, involving the choice synchronization structures of the initial net, the top down refinement of operations, the addition of resource places to form parallel and sequential mutual exclusions and the use of the token capacity to find the initial marking, as illustrated in this case study, guarantees the final net is live, bounded, and reversible. This is significant because it avoids the necessity for analysis for these properties. The authors observed that while top-down design for operations is relatively easily done, the hierarchical representation of shared resources is very difficult for this system.

The final Petri net model and the interpretation of the places for this FMS are shown in the Fig. 7.6 and Table 7.3. There are a total of 68 transitions which represents the start or completion of the related operations. The initial tokens in P_1,P_6, and P_7 have been specified, i.e., $2 \leq m_0(P_1) \leq 20$, $1 \leq m_0(P_6)$, and $1 \leq m_0(P_7)$. Each place P_3, P_{10}, P_{16}-19, P_{31}, P_{37}, P_{39}, P_{41}, P_{60} has one token. $m_0(P_{28}) = 19$. The other places initially have no token.

The complexity of this net, as observed from Fig. 7.6, derives from the number of shared resources with which this FMS must cope. Despite this graphical complexity, one can still visualize many discrete event control characteristics from the net: choices, sequential relations, concurrency, mutual exclusions, and synchronization. With this net one can trace the sequences of

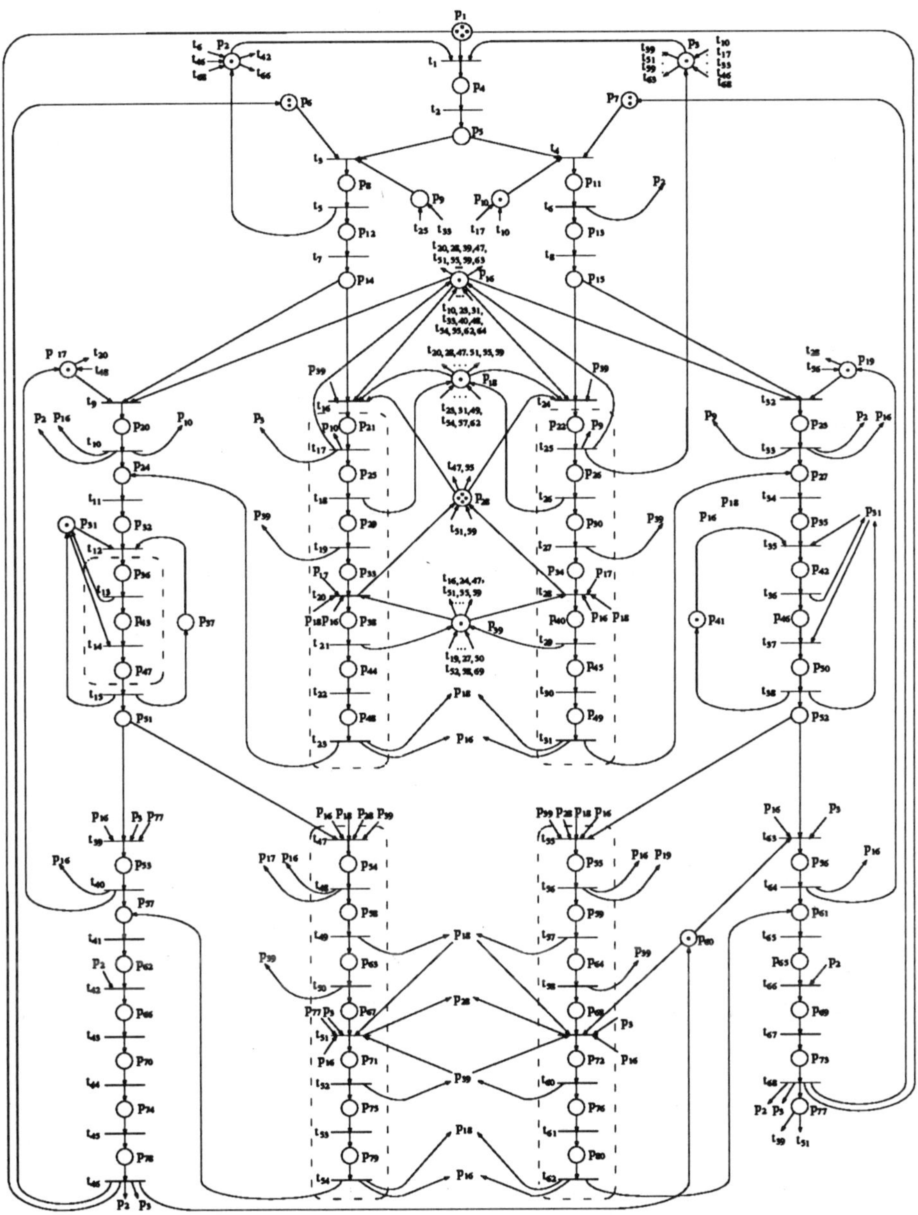

Figure 7.6: A final Petri net model for the FMS

Table 7.3: Place labels for the final Petri net

P_1: Pallet available
P_2: Scorbot robot available
P_3: C4 available
P_4: Move a pallet from the storage to C4
P_5: Empty pallet on C4
P_6: Raw pegs available
P_7: Raw blocks available
P_8: Move a raw peg from storage to pallet on C4
P_9: Get a peg second
P_{10}: Get a block first
P_{11}: Move a raw block from the storage to the pallet on C4
P_{12}: Move a peg-carrying pallet down C4
P_{13}: Move a block-carrying pallet down C4
P_{14}: A peg-carrying pallet on C4 available
P_{15}: A block-carrying pallet on C4 available
P_{16}: Gantry robot available
P_{17}: C3 available
P_{18}: C1 available
P_{19}: C2 available
P_{20}: Move a raw-peg-carrying pallet from C4 to C3
P_{21}: Move a raw-peg-carrying pallet from C4 to C1
P_{22}: Move a raw-block-carrying pallet from C4 to C1
P_{23}: Move a raw-block-carrying pallet from C4 to C2
P_{24}: A raw-peg-carrying pallet on C3
P_{25}: Move a raw-peg-carrying pallet up C1 and elevator into the ASRS
P_{26}: Move a raw-block-carrying pallet up C1 and elevator into the ASRS
P_{27}: A raw-block-carrying pallet on C2
P_{28}: The ASRS has empty slots
P_{29}: Store a raw-peg-carrying pallet in ASRS
P_{30}: Store a raw-block-carrying pallet in ASRS
P_{31}: Microbot robot available (note P_{31} appears twice on Fig. 7.6)
P_{32}: Move a raw-peg-carrying pallet up C3 to Lathe
P_{33}: Raw-peg-carrying pallets available
P_{34}: Raw-block-carrying pallets available
P_{35}: Move a raw-peg-carrying pallet up C2 to Mill
P_{36}: Load Lathe with raw peg
P_{37}: Lathe available
P_{38}: Retrieve a raw peg from ASRS
P_{39}: The ASRS elevator available
P_{40}: Retrieve a raw peg from ASRS
P_{41}: Mill available
P_{42}: Load the Mill with a raw block
P_{43}: Machining a raw peg
P_{44}: Move a raw-peg-carrying pallet down C1
P_{45}: Move a raw-block-carrying pallet down C1
P_{46}: Machine raw block

Table 7.3: (Continued)

P_{47}:	Unload finished peg to pallet on C3
P_{48}:	Move raw-peg-carrying pallet from C1 to C3
P_{49}:	Move raw-block-carrying pallet from C1 to C2
P_{50}:	Unload finished block to pallet on C2
P_{51}:	Move finished-peg-carrying pallet down C3
P_{52}:	Move finished-block-carrying pallet down C2
P_{53}:	Move finished-peg-carrying pallet from C3 to C4
P_{54}:	Move raw-peg-carrying pallet from C3 to C1
P_{55}:	Move raw-block-carrying pallet from C2 to C1
P_{56}:	Move finished-block-carrying pallet from C2 to C4
P_{57}:	Finished-peg-carrying pallet on C4 available
P_{58}:	Move finished-peg-carrying pallet up C1 and the elevator into the ASRS
P_{59}:	Move finished-block-carrying pallet up C1 and the elevator into the ASRS
P_{60}:	Assembly station empty
P_{61}:	A finished-block-carrying pallet on C4 available
P_{62}:	Move a finished-peg-carrying pallet up C4
P_{63}:	Store finished-peg-carrying pallet in ASRS
P_{64}:	Store finished-block-carrying pallet in ASRS
P_{65}:	Move finished-block-carrying pallet up C4
P_{66}:	Move finished peg into the Assembly Station
P_{67}:	Finished-peg-carrying pallets in ASRS available
P_{68}:	Finished-block-carrying pallets in ASRS available
P_{69}:	Place finished block in Assembly Station
P_{70}:	Assemble peg-block product
P_{71}:	Retrieve finished peg from ASRS
P_{72}:	Retrieve finished block from ASRS
P_{73}:	Put pallet from finished-block-carrying pallet in pallet storage
P_{74}:	Move assembled product to output storage carousel
P_{75}:	Move finished-peg-carrying pallet down C1
P_{76}:	Move finished-block-carrying pallet down C1
P_{77}:	Finished block in the Assembly Station
P_{78}:	Put pallet from a finished-peg-carrying pallet in the pallet storage
P_{79}:	Move finished-peg-carrying pallet from C1 to C3
P_{80}:	Move finished-block-carrying pallet from C1 to C2

operations and all the required resources for any transitions to fire (initiation of an operation). The graphical complexity is partially hidden because some arcs from resource places to transitions are not completely drawn In addition, one can observe which operations share a resource.

7.4 Implementation of the Discrete Event Control

The implementation method used is to code the net using a Petri net description language and to supervise the system by employing a Petri net token player which uses the net description as data.

7.4.1 Petri Net Description Language (PNDL)

The Petri net description language is a declarative language which is developed and used to specify a Petri net. Using such a language, the graphical notation of a Petri net can be converted to a text representation. Each PNDL text file defines information about places, transitions, and the connectivity of the Petri net model. The input to the system is in the form of declarations of the node lists (places and transitions) by type and a declaration of attributes for each place and transition.

A node list declaration specifies a type followed by a list of one or more identifiers of that type. The identifiers in PNDL are the names of places and transitions. The identifiers are the user defined labels, e.g., T_1, of the places and transitions. The types are: *pl-simple, pl-action*, and *trans*. *pl-simple* declares places to represent resources or buffers and *pl-action* declares them to represent operations. All transitions are of type *trans*, for example, *trans* T_1, T_2, T_3.

An attribute declaration consists of the node identifier followed by a group of attribute types enclosed by braces. Each specified attribute is followed by a list of one or more node identifiers, integers or character strings. Examples of these attribute types are as follows:

descr: unique character string describing the place or the transition;

inp: identifier list of input places of a transition;

outp: identifier list of output places to a transition; and

mark: integer number of initial tokens marking a place.

For example, $T_1\{$***inp*** p_1, p_2; ***outp*** p_4; ***descr*** "Finish assembling"$\}$, means that transition T_1 has input places p_1 and p_2, output place p_4, and represents "Finish assembling". These declaration are processed into place and transition files. Note that each declaration requires only information local to the place or transition. This makes for relatively straightforward net coding. The reader is referred to [Rudolph 89] for more detail.

7.4.2 Petri Net Supervisory System Commands

A Petri net supervisory system is designed to execute the basic tasks of a system supervisor: to initiate and coordinate operations between workstations to perform a desired task. PNSS interprets the information contained in the Petri net model files and communicates with workstation controllers to initiate appropriate operations through the workstation computers. It consists of a Petri net execution algorithm and a set of PNSS commands. The commands are invoked to interpret a Petri net model described in a PNDL text file. They can also be used to interactively execute a Petri net model and to display information about the marking of the net. These commands are summarized as follows:

LOAD This command is used to interpret a PNDL file. PNSS checks the load file for correct syntax and initializes the necessary place and transition data structures. When the command is invoked and no errors occur, the Petri net model is loaded into memory. When PNSS detects an error in the PNDL file, it prints a message indicating the cause of the error.

RUN This command begins execution of the PN model whose PNDL file has been interpreted with the LOAD command. Execution causes transitions to fire. PNSS responses indicate which places are being marked and unmarked and what operations, if any, are in progress at workstations. Executions may be halted for reasons such as:

- A breakpoint places (set by users) becomes marked.
- User interrupts by entering $'p'$ or $'P'$ on the keyboard.
- The net is no longer live.

STN It indicates what workstations a Petri net model will utilize. It should be used before the model that contains action or operation places is run.

BPL A user can use BPL to set breakpoints at places. Execution is halted when a place with a breakpoint is marked.

BC It reverses the effect of the BPL command by clearing breakpoints. BPL and BC are helpful in debugging a PN model.

CONT It is used to resume execution of a PN model after a breakpoint.

TOFF It "turns off" selected transitions. Transitions that are off cannot be enabled.

TON It reverses the effect of TOFF by "turning on" the selected transitions.

P and T These two commands are used to display information about the selected places and transitions, respectively, when the execution of a PN model is suspended. The information for a place includes type, breakpoint status, marking, input/output transitions, etc. and input/output places, on/off status, etc. for a transition.

RES This command is used to set conflict resolution strategies for enabled transitions.

LOG It is used to have the responses from PNSS during the execution of a PN model saved in a disk file.

EXIT It is used to end the PNSS session and return to MS-DOS.

7.4.3 Petri Net Execution Algorithm

The kernel of the PNSS is a token player program, i.e., the Petri net execution algorithm, which executes any net described by the data files. Written in the C programming language, PNSS runs on an IBM PC-XT computer and is used to successfully control the concurrent operations of the described FMS. The execution algorithm is explained as follows and its flow diagram is depicted in Fig. 7.7.

The initialization includes setting the newly marked list of places (L) as the initially marked places, and the list of enabled transitions (E) empty. The elements in L will be ordered as a First-In-First-Out (FIFO) queue. As shown in Fig. 7.7, the first place in L is removed from L. If it is a resource place, its output transitions are checked to see if they are enabled. If none of its output transitions is enabled, it will be placed at the end of L. If any output transition is enabled, it is added to E. If there is any enabled transition, i.e., E is not empty, then one is selected to fire. Firing of a transition consists of unmarking its input places and removing them from L, marking its output places and adding them to L.

If the place type is operation as opposed to resource, it is checked to see if it is set to *Phase-Start*. If the place is in Phase-Start, the program will request an operation at the appropriate workstation, as depicted in Figs. 7.7 and 7.8, and will then poll all workstations for completed operations. Each place set to Phase-Start will remain set until a completion signal is returned

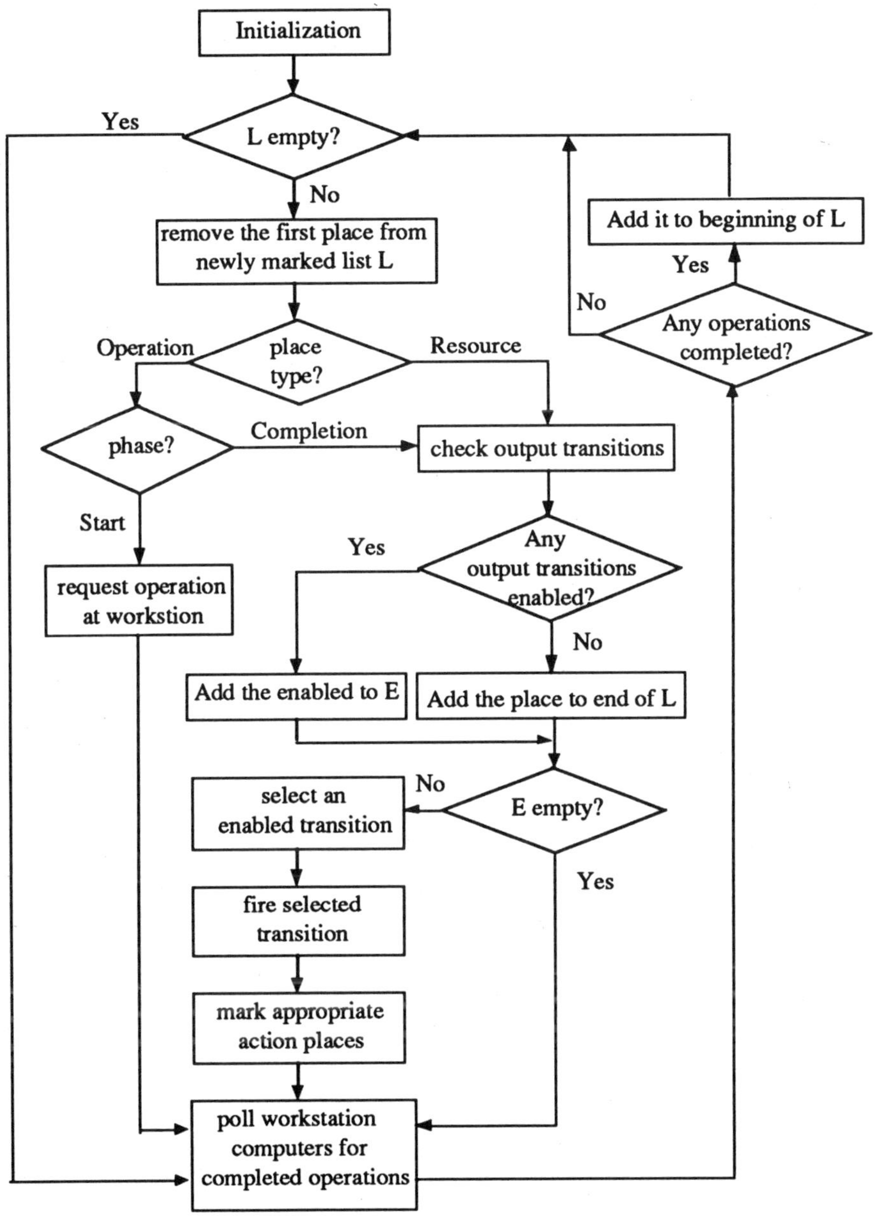

Figure 7.7: The Petri net execution algorithm

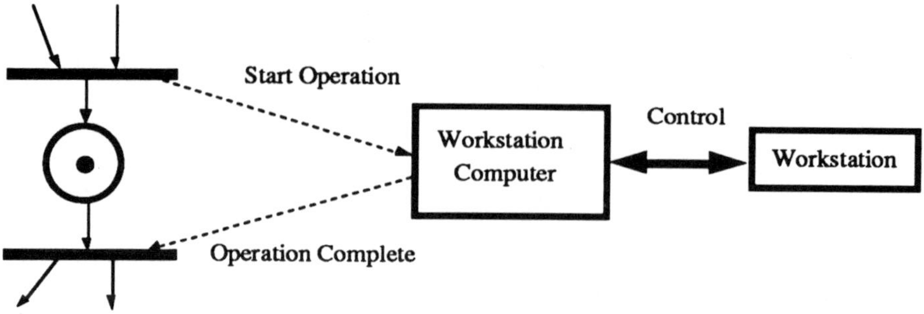

Figure 7.8: Interface between the PN supervisor and a workstation computer

by its workstation computer. In each iteration all operations are checked for completion and the places representing operations which are completed are set to *Phase-Completion* and are added to the beginning of L. When $L = \emptyset$, the algorithm checks the workstation computers repeatedly. The change in status from Phase-Start to Phase-Completion allows a marked place to enable its output transitions. If the operation place is set to Phase-Completion, the algorithm then checks its output transitions and so on as in the case of a resource place as discussed above.

For the case of more than one transition enabled at a time, there can be any number of selection mechanisms for which to fire, for example, a First-In-First-Out (FIFO) policy and a priority policy. The details of implemented selection mechanisms and software documentation can be found in [Rudolph 89].

The PNSS code can be summarized as the following major functional components:

1. The PNSS command shell which interprets the PNSS commands and invokes command execution code.

2. The interpreter code interprets PNDL files and initializes the necessary place and transition data structures. The code is invoked by the LOAD command.

3. The execution code implements the PN execution algorithm and other routines to alter execution (breakpoints and conflict resolution) and determine the status of the PN models. It is invoked by RUN and CONT, and stopped by EXIT.

4. The communication code includes routines to handle communications between PNSS and workstation computers.

Communication and discrete event control are implemented using a host computer, an IBM PC-XT, called the system supervisor in this FMS, together with six Apple computers. These Apple computers are mainly used to control the local workstations as shown in Fig. 7.2. The FMS modeled here has been realized in the Industrial Automation Laboratory at Rensselaer Polytechnic Institute with the help of the Petri net supervisory system described above.

7.5 Summary

The hybrid synthesis methodology has been successfully applied to the design and implementation of a Petri net model for a flexible manufacturing system of moderate complexity. Using top-down stepwise refinement and bottom-up composition with mutual exclusion theory, a Petri net has been systematically constructed such that it exhibits system properties of boundedness, liveness, and reversibility. This avoids the necessity for complex analysis for these properties of this complicated Petri net. The resulting Petri net model has been used to generate a supervisory controller for the existing physical FMS simulator. The implementation of the Petri net as a supervisor is briefly discussed by describing a Petri net description language and a Petri net execution algorithm.

The system, operating in a laboratory, typifies the processes and resources found in a factory shop floor. It illustrates what researchers and practitioners realize theoretically and intuitively; that the increased flexibility in automation provided by interacting programmable processes complicates the control design necessary for valid and efficient operation of such systems.

The significance is that it shows that the current state of the art in Petri net theory provides for a tractable methodology to develop sequence control for a complex FMS. It further illustrates that fairly straightforward computer tools can be developed to encode complex Petri nets and subsequently execute them to control the physical system.

Chapter 8

Augmentation of Petri Nets

8.1 Introduction

The research into augmentation methods of Petri nets is motivated by the desire to improve control coding for error handling in automated systems [Zhou 89a]. In implemented automated manufacturing workstations or cells, an inordinate amount, up to 90% [Gini 83; Lee 83; Srinivas 78; Fielding 88] , of the control coding effort is dedicated to exception handling or automatic error recovery. At present, most of this coding effort occurs at the design stage. Engineers attempt to anticipate common errors and write the control code to handle them and to allow the system to automatically recover if feasible. Two undesirable outcomes are possible. First, in an effort to be thorough, routines are coded to handle errors which in fact do not occur during the lifetime of the program. Second, errors that have not been anticipated will inevitably occur when running the system and the code may have to be "patched" in order to handle future occurrences of the same errors. Changing the code after the design phase presents many potential software maintenance problems not the least of which is the possibility of the introduction of coding errors. The results in this chapter allows for augmenting, i.e., patching, the control code while minimizing the possibility of introducing new errors. This will provide for a coding strategy where only the most common errors are accounted for at system design stage and the remainder are treated as they occur while the system is running. Many errors will occur during the initial system burn-in and can be treated at that time. This may alleviate the first undesirable property.

This chapter assumes that the given control code has been specified as a Petri net. When control code is specified in this way we call it a Petri net controller because the Petri net representation can be compiled and downloaded into a programmable logic controller (PLC) or a control processor using an algorithm as demonstrated in Chapter 7. This chapter presents a methodology which allows the augmentation of control code for error recovery while preserving the behavioral properties of the logic to avoid deadlock, buffer overflow and other characteristics as discussed in Chapter 2. If the changes in the code are made as prescribed here then the properties are guaranteed without further analysis. The magnitude of the analysis effort is so large that it is worth avoiding, particularly for complex systems.

Thus, the goal of this chapter is to show how we can augment an existing Petri net for the purpose of error recovery and still preserve the desirable properties guaranteed in its initial design. Our objectives are as follows:

1. To propose the concept of a Petri net controller for automatic error recovery and investigate its design in the context of automated manufacturing;

2. To study four basic augmentation methods for different types of error recovery;

3. To prove that the behavioral properties of the controller are guaranteed when the four basic Petri net augmentation methods are applied; and

4. To use a robot assembly cell as an example to demonstrate the proposed approach.

An architecture for an intelligent automatic error recovery system is shown in Fig. 8.1, where both a Petri net controller and machine intelligence techniques are used. The latter aims mainly at error diagnosis [Chang 91] and recovery planning [Kumaradjaja 89] . The Petri net controller reacts to detected errors by changing the control flow of tokens in the Petri net during the operation of the system. The controller includes the use of sensory data and machine intelligence for error diagnosis and recovery planning. This chapter focuses on the Petri net controller.

8.2 Design of Petri Net Controllers

8.2.1 Background

The token player in Petri nets is well adapted to execute the normal functions of a system [Atabakche 86] . Therefore, our idea is to attach some information

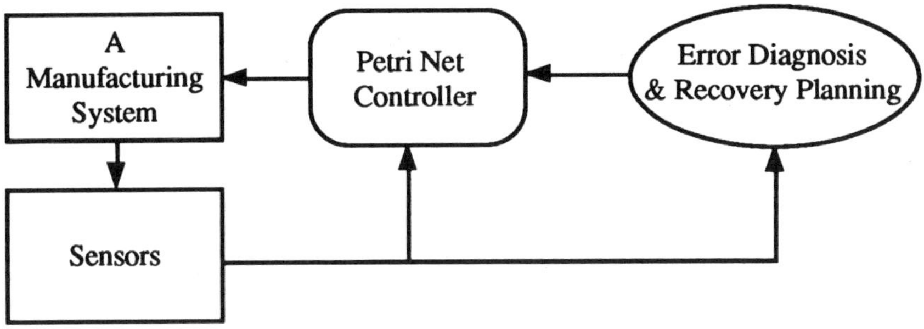

Figure 8.1: An intelligent automatic error recovery system

to places and/or transitions in a Petri net so that *abnormal* states of the system will be processed using a Petri net controller. When the system works normally, the sequences represented in the original Petri net will be executed. Otherwise, as an error occurs, a special part of the Petri net will be started depending on the type of error that has occurred.

In our system, there are two approaches to indicating the occurrence of an error. One is based on the information from the sensors of the manufacturing system. The other is to use watchdog timers in the Petri net controller. The use of watchdog timers is motivated by the previous work in communication protocols [Merlin 76a] . These timers indicate a possible error occurrence if a token stays in the place beyond a given time.

8.2.2 Description of Petri Net Controllers

A Petri net controller is defined as a control logic based on a marked Petri net for a discrete event dynamic system. Our approach to constructing a Petri net model for a system is to use places rather than transitions to model system processes as discussed in Section 2.3. The advantage is that it preserves the convention of transitions as instantaneous events, e.g. the start and completion of process execution. The presence of a token in a place indicates the activity of a process, e.g., pick or move, or a state of a device, e.g., the working state of a machine. This approach has been shown to easily lead to the stepwise refinement of a Petri net in Chapter 5; thus, a hierarchical Petri net controller can be constructed.

8.2.3 ε-function, τ-function, and a Constant

In implementing a Petri net controller for a real-time system, time requirements need to be satisfied. Therefore, it is necessary to introduce the time

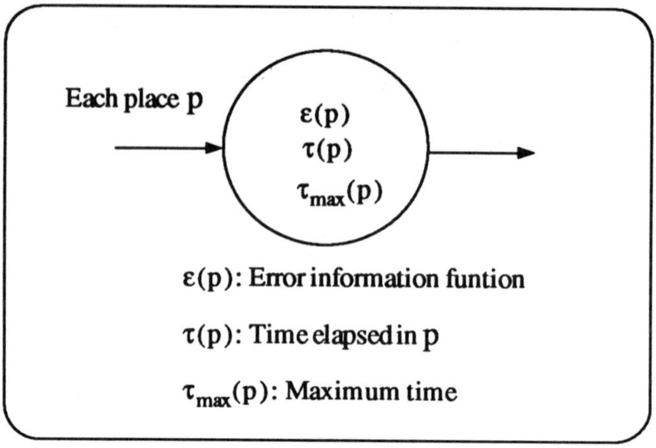

Figure 8.2: A place with $\varepsilon(p), \tau(p)$, and $\tau_{max}(p)$ in a Petri net

variable to the Petri net. In our controller, time is required only in operation places which represent the processes or operations. In order to process *abnormal* states that may appear during the execution of the system and detect errors using this Petri net, two functions and a constant are attached to each place in P (see Fig. 8.2):

$\varepsilon : P \mapsto N = \{0, 1, ...\}$ and is called the error information function (also called the ε-function);

$\tau : P \mapsto R = \{x, x$ is a non-negative real number$\}$ and $\tau(p)$ (called the τ -function) is the elapsed time in a place p in P; and

$\tau_{max} : P \mapsto R$, and $\tau_{max}(p)$ is the maximum elapsed time in a place p.

These are explained as follows:

1. The elapsed and maximum times: $\tau(p)$ and $\tau_{max}(p), \forall p \in P$

   $\tau_{max}(p)$ is the maximum amount of time during which a token can stay in a place p. $\tau(p) > \tau_{max}(p)$ implies that an error may have occurred in the system so that the mechanism of fault diagnosis will be started.

2. The error information function: $\varepsilon(p), \forall p \in P$

   $\varepsilon(p) = 0$: the process or state that place p represents is being executed if $\tau(p) \leq \tau_{max}(p)$. Otherwise, it is dead. This implies that some error may have occurred.

$\varepsilon(p) = 1$: the normal execution of the Petri net takes place if there is at least one token in the place p. Initially, $\varepsilon(p) = 1, \forall p$ in the Petri net controller.

$\varepsilon(p) = 2$: emergency shut down procedure starts if there is a token in this place p.

$\varepsilon(p) \leq 3$: the $(\varepsilon(p) - 2)$nd error is handled by executing a special part of the Petri net controller.

8.2.4 Adaptive Design of Petri Net Controllers

A Petri net controller can be easily extended for real-time purposes as we discussed in the last chapter. In the following discussion, a Petri net controller is one based on $Z = (P, T, I, O, m_0, \varepsilon, \tau, \tau_{max})$ where ε, τ, and τ_{max} are defined in the last subsection. The controller, denoted by Z, can be implemented for operation of the system. The initial value of $\varepsilon(p)$ is set equal to 1 for every $p \in P$. The adaptive design process for the Petri net controller can be described as follows:

1. If no error takes place or an old error is detected through the sensor data then Z will be executed without augmentation of this Petri net controller;

2. If a new error occurs, the mechanism of fault diagnosis and error recovery planning will be started by using, for example, machine intelligence techniques or operator intervention. The result of this error recovery planning will be expressed as a part of a Petri net which will be added to this controller. This results in a new Petri net controller Z'. The system will recover from the error under the control of this new Petri net controller Z'.

The controller Z or later Z' changes when any new error takes place. The construction of the new controller Z' from the old controller Z and the error-handling procedure S' is shown in Fig. 8.3. S' consists of place set P_S, transition set T_S, input function I_S, output function O_S, and its initial marking m'_0. Mathematically, the new controller can be expressed as the following:

$$Z' = (P', T', I', O', m', \varepsilon', \tau', \tau'_{max})$$

with two functions ε' and τ' and a constant τ'_{max} in each p in P', where

1. P' consists of P and all new places added to this Petri net controller or $P' = P \cup P_S$;

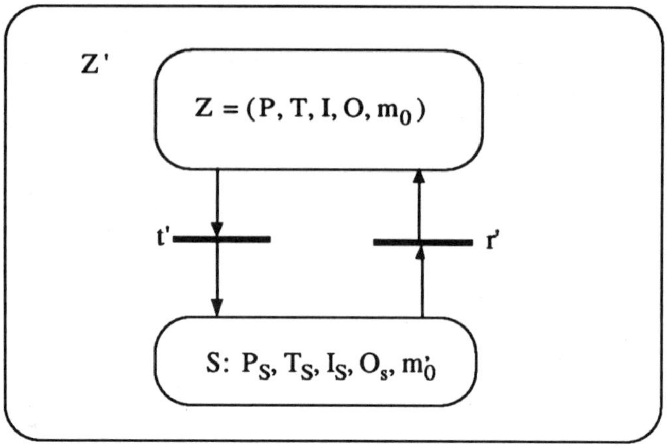

Figure 8.3: The Construction of the new Petri net controller Z' from the old controller Z and the error-handling procedure S' through transitions t' and r'

2. T' consists of T and all new transitions added to this controller or $T' = T \cup T_S \cup \{t', r'\}$;
3. I': $P' \times T' \mapsto \{0, 1\}$;
4. O': $P' \times T' \mapsto \{0, 1\}$;
5. m': $P' \mapsto \{0, 1, 2, ..., M\}$.
6. ε': $P' \mapsto N$ and is extended to each new place;
7. $\tau'(p)$ and $\tau'_{max}(p) : P' \mapsto R$.

I', O', and m' are extensions from I, O, m, respectively. They can be easily determined by Z, the error recovery block, and the relations between them (four basic augmentation methods are referred to the next section). The marking for the new controller Z' can also be readily obtained.

The firing rules for the Petri net controller in a real-time system are as follows:

1. A transition is enabled if and only if $m'(p) \leq 1, \varepsilon'(p) > 0$, and $\tau'(p) \leq \tau'_{max}(p)$ for each p that is one of its input places in Z'.
2. If a transition $t \in T'$ fires, $m''(p) = m'(p) + O'(p,t) - I'(p,t)$, for each $p \in P'$. $\varepsilon'(p) = 0, \tau'(p) = 0$, for each output operation place of the

transition t and the time starts counting for $\tau'(p)$ starts. When the process is completed normally, $\varepsilon'(p) = 1$. If an error is detected, it will be set to a particular value.

8.3 Augmentation Methods of Petri Nets

We desire to incorporate the results of error recovery planning to augment the controller Z to form Z'. Motivated by research done by Fielding, DiCesare and Goldbogen [1988, 1989], four basic methods for augmentation to Z' from Z are proposed. They are input conditioning, alternate path, backward error recovery, and forward error recovery, all of which are similar to the four methods discussed in [Fielding 88]. The last subsection will deal with the maintainability of this controller's behavioral properties through four theorems.

8.3.1 Input Conditioning Method

The idea of the input conditioning method is that an abnormal state in a place that represents a process or a state of a manufacturing system can become a normal state after other actions are finished or some conditions are satisfied. An example [Fielding 88] of this is the arrival of a part that requires additional finishing prior to processing at a workstation. Finishing that part prior to normal processing is an example of an input conditioning form of error recovery.

Figure 8.4 depicts a general augmentation to Z' from Z using the input conditioning method. Here Z is the original Petri net with place set P and transition set T; S' is a Petri net block which represents the error recovery procedure; T_S is the set of transitions in S'; P_S is the set of places in S'. They satisfy the conditions: p is a safe place, $P_S \cup T_S \neq \emptyset, P_S \cap T_S = \emptyset$, and $(P_S \cup T_S) \cap (P \cup T) = \emptyset$. The theorems of the last subsection show that such an augmentation will not change the liveness, boundedness (or safeness), or reversibility properties of Z given that S' satisfies certain conditions.

Two special cases for the input conditioning method are as follows:

1. Retry the execution of the process that this place represents (Fig. 8.5). For example, only some parameters in the execution of the same operation in a system need minor changing. A specific example is that when a robot fails to insert a part in a workpiece, the robot can retry to insert the part.

2. Before restarting this place, another action will be executed first (Fig. 8.6). For example, suppose a tool is broken. After the tool is changed, the

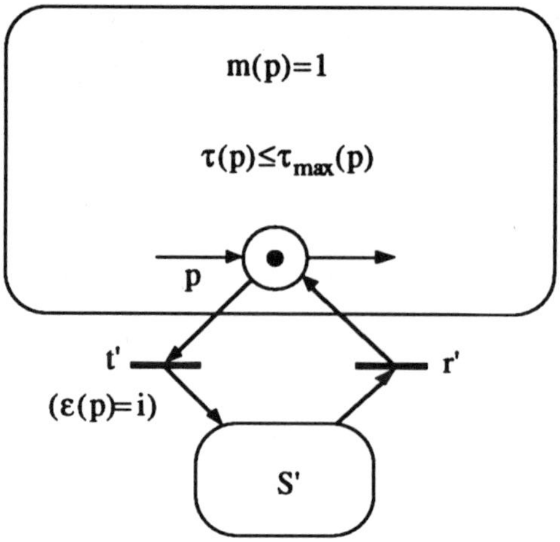

Figure 8.4: Input conditioning (restart p after running a Petri net S') where t' fires when $\varepsilon(p) = i$ and S' will be executed

machine is again in normal state. This constitutes error recovery using the input conditioning method.

8.3.2 Alternate Path Method

The philosophy of the alternate path method (also called error avoidance) states that there exists another sub-Petri net controller which can transform an abnormal state in place p directly into a normal state in a system. Figure 8.7 depicts this type of construction of Z' from Z. It is assumed that the state in the place w will be achieved after the state in the place p. S' and Z satisfy the same conditions described as in the last subsection. An example showing the alternate path method is cited as follows [Fielding 88, Zhou 88]. Two parts are fed to a robotic assembly cell in swapped feed positions. There is a feed pallet such that Parts 1 (2) are to be found at pallet positions 1 (2). A robot manipulator is used to retrieve Parts 1 (2) from pallet positions 1 (2) and install them at assembly positions A1 (A2). Now assume that Parts 1 (2) are found at pallet positions 2 (1) and the parts can be sensed as being in the swapped pallet positions. Then this error can be avoided if the robot manipulator can retrieve the part from pallet position 2 and install it at assembly position A1 and the other part from pallet position 1 and to install it at A2. Similarly, the last subsection gives proofs that the augmented net Z' will maintain the behavioral properties of Z given that S' has certain

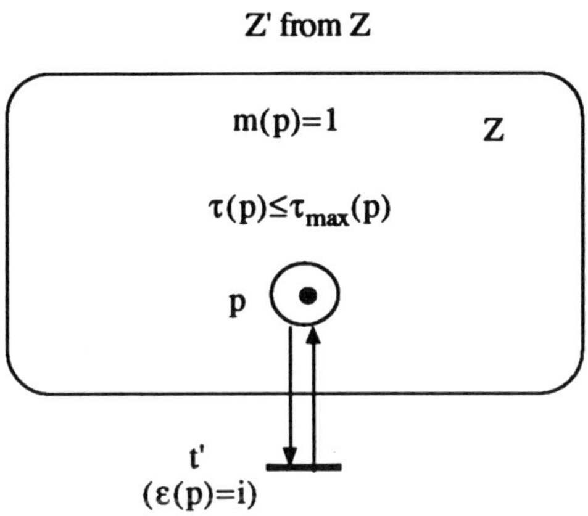

Figure 8.5: Restart place p (a special case of input conditioning method), i.e., fire t' when $\varepsilon(p) = i$

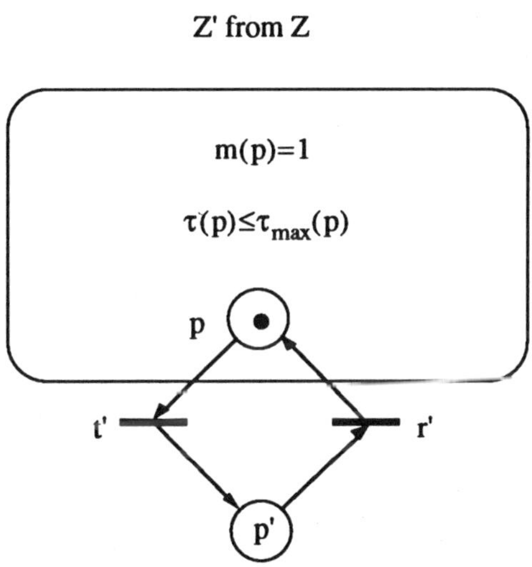

Figure 8.6: Restart place p after the process represented in the place p' (a special case of input conditioning method) where t' fires when $\varepsilon(p) = j$ and r' fires if process p' succeeds

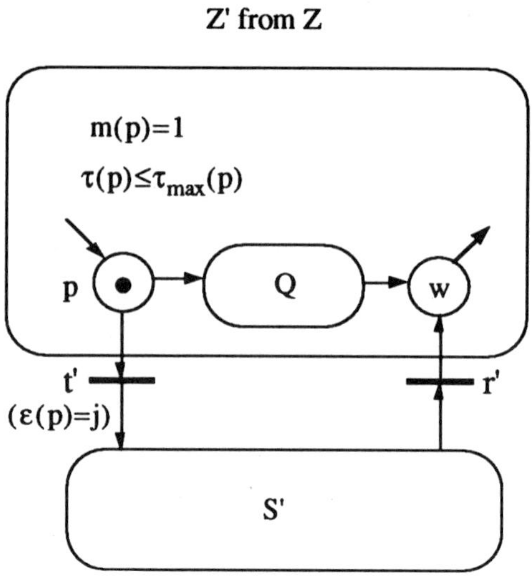

Figure 8.7: Alternate path method (substituting Q with S') where t' fires when $\varepsilon(p) = j$

properties.

8.3.3 Backward Error Recovery Method

Backward error recovery suggests that under the assumption that the state is normal, i.e. $\varepsilon(p) = 1$, Q in Z is executed and a new faulty state in place w results. But this state can become a normal state in place p after the operation of a sub-Petri net controller S'. This kind of construction of Z' from Z is shown in Fig. 8.8. Also, the same conditions about Z and S' are assumed. A typical example is given by Wiendahl, *et al.* [1985] as follows:

In the vibratory bowl feeder it is impossible to avoid the lack of order in parts. One cannot sense whether the screws in the bowl are correctly aligned so that they will not jam as they pass from the bowl to the feeder track. However, it is relatively easy to sense a jam in the feeder. If the jam takes place at a particular position, an air-driven plunger is employed to automatically push the screws back into the bowl, at which point the feed process is restarted.

Again, the desired properties are preserved, as shown by the theorems of the last subsection whenthi error recovery method is applied.

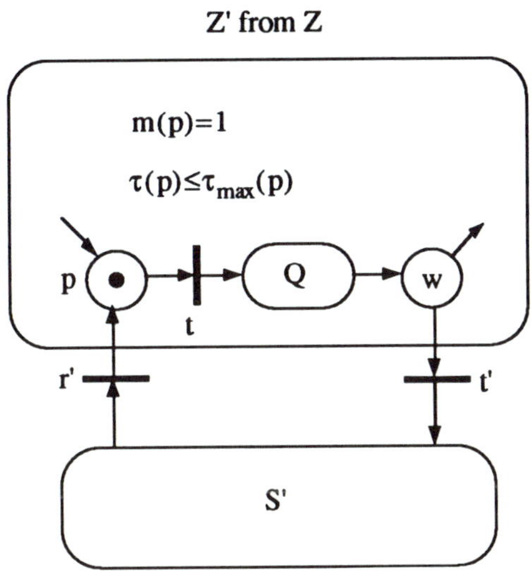

Figure 8.8: Backward error recovery method (starting Q as if it were normal) where t could fire assuming $\varepsilon(p) = 1$

8.3.4 Forward Error Recovery Method

The forward error recovery method is similar to the backward error recovery method; suppose that a faulty state results after the operation of Q in Z. However, this state can be directly transformed into a normal state in Z after a Petri net controller S' operates. Such a construction of Z' from Z is depicted in Fig. 8.9. Again, the same conditions about Z and S' are assumed. This strategy is particularly useful when a fault at p is difficult to detect while it can be relatively easier to detect at w. A simple example is cited from the paper [Feicht 87].

Assume that gears are retrieved from a feed pallet and assembled in a way that requires the gears to mesh. It is impossible to sense in advance if the gears will mesh or not when placed at their assembly positions. However, it is straightforward to determine that the gears haven't meshed after they are put at their assembly positions. By giving a twist until meshing occurs, the system can directly transform the situation into a normal state. The behavioral properties of this augmentation will also be preserved as shown in the next subsection. The results that unify these four methods into one structure are given by Fielding [1989].

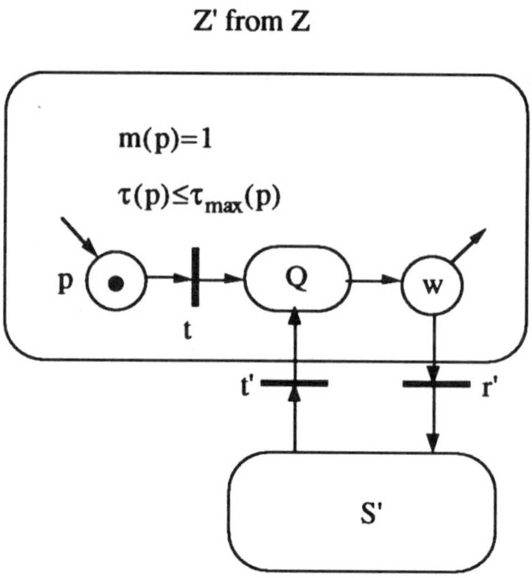

Figure 8.9: Forward error recovery method (starting Q as it were normal) where t could fire assuming $\varepsilon(p) = 1$

8.3.5 Maintainability of behavioral Properties

The property analysis of a Petri net includes boundedness, safeness, liveness (absence of system deadlock), and reversibility (re-initializability or properness) as defined in Chapter 2. The following will focus on the maintainability of these properties of a Petri net controller when Z' is constructed from Z using the four basic construction methods discussed above.

The four basic construction methods are input conditioning, alternate path, backward error recovery, and forward error recovery. For each method, there are t', r', S' that do not belong to the controller Z. Let S' be a Petri net block which represents the error recovery procedure with its initial marking m_0'.

An associated Petri net S^* is defined as a Petri net which consists of a place p, t', r', and S' where p is called an idle place [Valette 79]. Its initial marking is m_0*$= (1, m_0')$ where m_0' is the initial marking of S' and 1 implies that the idle place is initially marked. An example is shown in Fig. 8.10.

The idle place p in the associated Petri net S^* which represents an operation in the Petri net controller in an automated manufacturing system. At that operation place, some errors may occur and be detected so that error recovery is needed.

Let Z' be constructed from Z coupled with S', t', r', and two arcs (see

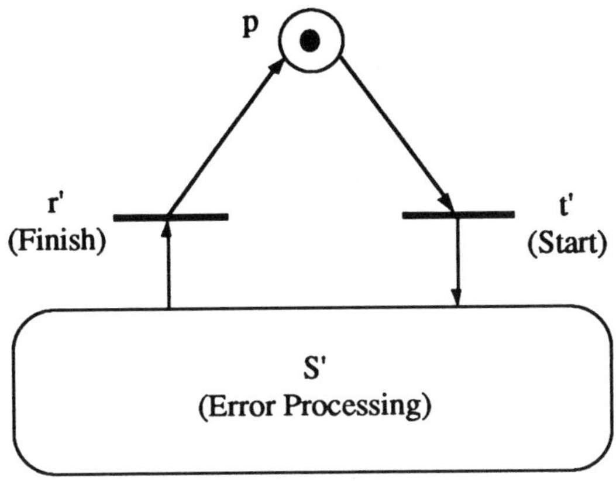

Figure 8.10: Associated Petri net S^* is defined by p, t', S', and r', and its initial marking $m_0{}^*= (1, m'_0)$ given the initial marking m'_0 of S' where "1" indicates that the p is marked

Figures 8.4, 8.7-9) by using the above four basic augmentation methods when an error occurs. We then have the following theorems.

Theorem 8.1 *Z', with respect to (m_0, m'_0), is bounded (or safe) if and only if both Z, with respect to its initial marking m_0, and S^*, with respect to its initial marking m^*_0, are bounded (or safe).*

Note that in following discussion, $R(Z, m_0)$ is the reachability set of Z with respect to m_0; $R(S^*, m^*_0)$ is the reachability set of S^* with respect to $m^*_0 = (1, m'_0)$; and $R(Z', M_0)$ is the reachability set of Z' with respect to $M_0 = (m_0, m'_0)$.

Assume that the place p is in the ith position in the n places of Z', where the first L places are from Z and the other $(n - L)$ places from S', $L \leq i$, and $n \geq L$. For any $M \in R(Z', M_0)$,

$$M = (M(p_1), ..., M(p_{i-1}), M(p), M(p_{i+1}), ..., M(p_L), M(p_{L+1}), ..., M(p_n))$$

denoted by

$$M = (m_1, m(p), m_2, m') = (m, m')$$

for short, where

$$m_1 = (M(p_1), M(p_2), ..., M(p_{i-1}))$$
$$m(p) = M(p)$$

$$m_2 = (M(p_{i+1}), ..., M(p_L), M(p_{L+1}), ..., M(p_n))$$

and

$$m = (m_1, m(p), m_2).$$

For $M_j \in R(Z', M_0), M_j$ is also written as $M_j = (m_{j1}, m(p), m_{j2}, m'_j)$. If g is a sequence of transition firings, then $M[g>$ shows that g is applied to marking M. If t is a transition, $M[gt>$ means that a sequence of transitions and transition t, in that order, are applied to the marking M.

Proof: The necessity is evident because any marking vector of S^* or Z is some part of a marking vector in $R(Z', M_0)$, and the latter is bounded or safe. The sufficiency is proved as follows. Assume that $\forall m \in R(Z, m_0)$, and $\forall m^* \in R(S^*, m_0^*)$, both m and m^* are bounded.

Consider that the input conditioning method is used.

Case 1. Given $M_0 = (m_0, m'_0), \forall M \in R(Z', M_0)$ that is reached after both transitions t' and r' fire K times ($K = 0, 1, 2, ...$) (Fig. 8.5). Then

$$M = (m, m') \text{ and } m \in R(Z, m_0), m^* = (1, m') \in R(S^*, m_0^*).$$

Therefore, M is bounded.

Case 2. M is reached after the latest firing of t' without firing r' correspondingly. But before the firing of t', $M_j = (m_{j1}, 1, m_{j2}, m'_j)$ has to be reached. Otherwise, t' cannot be enabled. $m = (m_{j1}, 1, m_{j2})$ is bounded. Only the firing of t' yields the marking

$$M_k = (m_{j1}, 0, m_{j2}, m'_k)$$

But m_{j1} and $1, m_{j2}$ may change into m_{k1} and $1, m_{k2}$, since some transitions in Z may fire; and m'_k may change into m'_v, since some transitions may fire in S' so that $M = (m_{k1}, 0, m_{k2}, m'_v)$. However, $(m_{k1}, 1, m_{k2}) \in R(Z, m_0)$ and $(0, m'_v) \in R(S^*, m_0^*)$. Since m_{k1}, m_{k2}, and m'_v are bounded, M is also bounded.

Therefore, $\forall M \in R(Z', M_0)$ is bounded.

Similarly, it can easily be shown that the same conclusion holds for the other three basic methods. Safeness is a special case of boundedness, i.e., 1-boundedness. Therefore, the theorem is proved. QED

Theorem 8.2 *Z' with respect to (m_0, m'_0) is live if and only if both Z, with respect to m_0, and S^*, with respect to $m_0{}^*$ are live.*

Proof: The necessity is evident because Z' would not be live if either S^* or Z is not live. The sufficiency is proved as follows. Consider that the first method, input conditioning, is used. For any transition t in $Z, \forall M \in R(Z', M_0)$, there are two cases:

Case A. t is a transition in Z but not one in S^*. Let $M = (m, m')$.

A-1. $m \in R(Z, m_0)$.

Since Z is live, $\exists$ a sequence of transition firings h in Z such that at marking $m[h>$, t is enabled. Apply h to M in Z', $M[h>$ producing $(m[h>, m')$ which will enable t.

A-2. $m \in R(Z, m_0)$.

Due to the construction of Z', this situation results only if a token in p leads t' to firing and p is no longer marked (Fig. 8.5). Therefore,

$$M = (m_1, M(p), m_2, m'), M(p) = 0.$$

But

$$(m_1, 1, m_2) \in R(Z, m_0)$$

and

$$(0, m') \in R(S^*, m_0^*).$$

Since S^* is live, $\exists$ a sequence of transition firings g such that r' is enabled. The result is

$$((0, m')[gr'> (1, m'').$$

Applying gr' to Z' at marking M,

$$M[gr' > (m_1, 1, m_2, m'')$$

Since $(m_1, 1, m_2) \in R(Z, m_0)$, this is case A-1.

Case B. t is a transition in S^* but not one in Z. Let $M = (m_1, m(p), m_2, m')$.

B-1. $(m(p), m') \in R(S^*, m_0^*)$.

Since S^* is live, $\exists$ a sequence of transition firings h such that at marking $(m(p), m')[h> (i, m'')$, t is enabled $(i \leq 0)$. Apply h to M in Z', $M[h> (m_1, i, m_2, m'')$ will enable t.

B-2. $(m(p), m') \in R(S^*, m_0^*)$.

Due to the construction of Z' and S^* (Fig. 8.5 and Fig. 8.9), $m(p) = 0, (1, m') \in R(S^*, m_0^*)$, and $(m_1, 0, m_2) \in R(Z, m_0)$.

Since Z is live, $\exists$ a sequence of transition firings h such that

$$(m_1, 0, m_2)[h> (m_a, 1, m_b).$$

Applying h to Z',

$$M[h> (m_a, 1, m_b, m').$$

But

$$(1, m') \in R(S^*, m_0^*).$$

This is case B-1.

Therefore, Z' is live with respect to $M_0 = (m_0, m'_0)$.

Similarly, we can easily prove that the conclusion holds true for the other three basic augmentation methods. QED

Theorem 8.3 *If S^* is reversible with respect to $m_0{}^*$ such that when the place p has a token, $m_0{}^*$ is reached, then Z' is reversible with respect to $M_0 = (m_0, m'_0)$ if and only if Z is reversible with respect to m_0.*

Proof: The necessity is obvious. The sufficiency is proved as follows. Consider that the first method, input conditioning, is used. $\forall M \in R(Z', M_0)$, there are two cases:

Case 1. $M = (m, m'_0), m \in R(Z, m_0)$.

Since Z is reversible, $\exists$ a sequence of transition firings h such that $m[h> m_0$. Apply h to M in Z', $M[h> (m_0, m'_0) = M_0$ is obtained

Case 2. $M = (m_1, M(p), m_2, m'), M(p) = 0$. (If $M(p) = 1$, then $m' = m'_0$, according to the given conditions). But $m^* = (0, m') \in R(S^*, m_0^*)$.

Since S^* is reversible, $\exists$ a sequence of transition firings g such that

$$(0, m')[g> m_0^* = (1, m'_0).$$

Apply g to M in Z',

$$M[g > (m_1, 1, m_2, m'_0) = (m, m'_0)$$

is obtained. But this is case 1. So $\exists h'$ such that

$$M[gh'> M_0.$$

From the given conditions, only the two above cases exist for any $M \in R(Z', M_0)$.

Therefore, Z' is reversible with respect to $M_0 = (m_0, m'_0)$.

Similarly, it can be proved that the conclusion holds true for the other three basic augmentation methods. QED

Before the next theorem is stated, the essentially decision free (EDF) place proposed by Krogh and Sreenivas [1987] will be introduced.

A place $p \in P$ is said to be *essentially decision-free (EDF)* with respect to an initial marking m_0 if $\forall m \in R(m_0)$ (reachability set), at most one output transition of p is enabled.

This concept generalizes the notion of decision-free Petri nets where each place has exactly one output transition. The EDF property states that no more than one of these transitions is enabled. In other words, when a place is EDF, the Petri net firing logic is such that there is never an ambiguity about which of its output transitions will fire.

Theorem 8.4 *The attachment of S' to any place p in Z does not change this place's EDF property.*

The proof of Theorem 8.4 is evident. This is because the error information function $\varepsilon(p)$ attached to p has its unique corresponding firing path when $\varepsilon(p)$ is determined.

The above four theorems imply that the behavioral properties of a controller remain if the four basic construction methods, input conditioning, alternate path, backward error recovery, and forward error recovery are employed.

8.4 An Application Example

8.4.1 Description

This example uses a piston stuffing workstation developed at the General Motors Technical Center. The workstation uses two robots to place and pull the pistons into the cylinders of an engine block and attaches the piston rods to the crank shaft. This system's layout is depicted in Fig. 8.11 and comprises the following components:

1. An indexing table with piston rods and caps;
2. A table with nut dispensers and piston pulling tools;
3. A workstation with an engine block;
4. A GMF S-380 robot;
5. A V-220 vision system;
6. A GMF M-1 robot; and
7. Some sensors installed in the robots.

The basic steps involved in the system are as follows:

1. The S-380 robot moves the crank shaft to align the crank shaft position for piston insertion;
2. The vision system checks and verifies the crank shaft position in the engine block;
3. The S-380 robot picks up a piston, with its rod attached, from the indexing table and brings it to an engine block cylinder; and

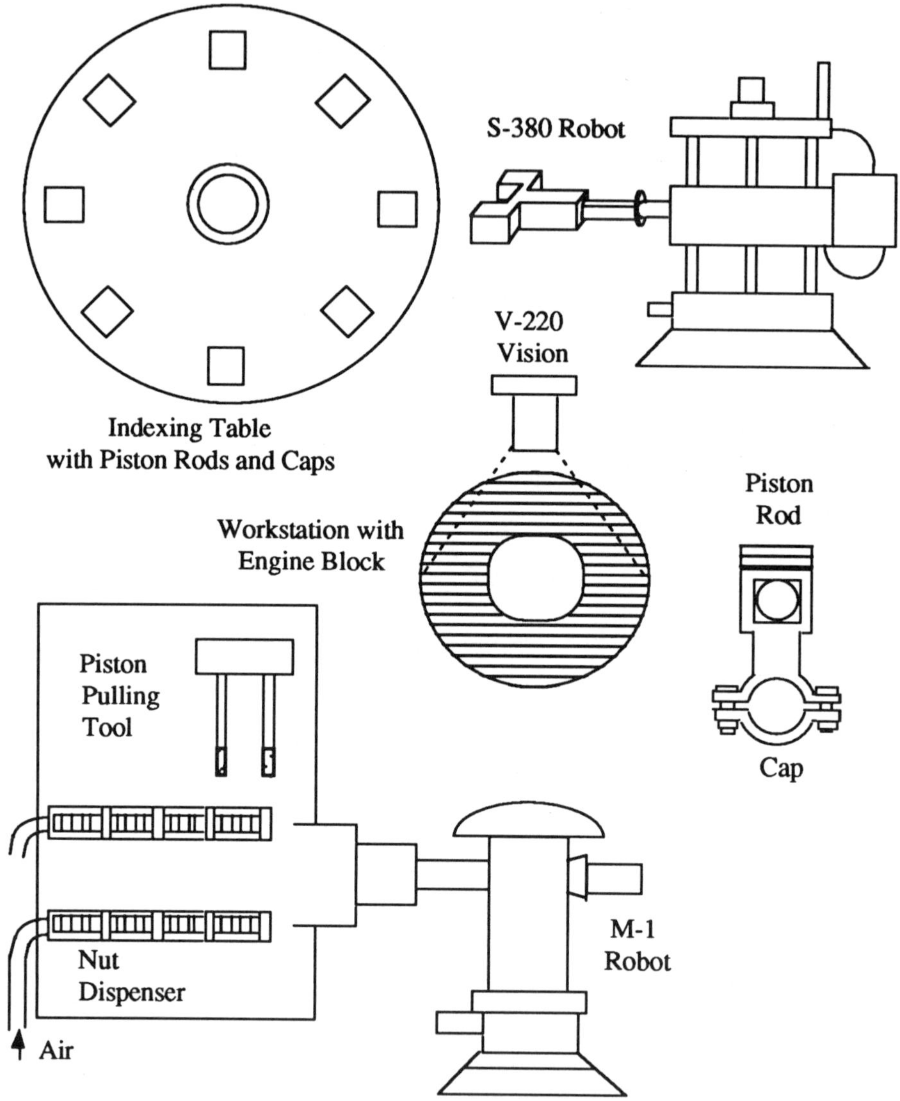

Figure 8.11: An overall layout of a robotic cell for piston assembly

4. The M-1 robot picks up the piston pulling tool, two nuts, and a cap and moves to the workstation, respectively. It pulls the piston rod into the engine block, installs the cap, and attaches and turns two nuts to secure this cap.

Repeat steps 1 through 4 for the remaining pistons.

Using the synthesis approach described in Chapter 5, the initial Petri net controller for accomplishing the above tasks can be obtained as shown in Fig. 8.12 using its first three steps or top-down design stage [Zhou 89b]. Note that the net is a marked graph. Therefore, Murata's marked graph synthesis methods [1980a, b] can be used to synthesize and to guarantee the properties of the resulting net.

This net has the desirable properties, safeness, liveness, and reversibility. It is also essentially decision-free; in fact, it is choice-free with respect to its initial marking

$$(1,1,1,0,1,1,1,0,0,0,0,1,0,0,1,0,0)^{\tau}.$$

There are 17 safe places which represent the availability of resources or the operations and 9 transitions which represent the the start and/or completion of operations as described in Fig. 8.12 and Table 8.1.

8.4.2 Error Recovery

As an example of an input conditioning error recovery, suppose that the piston puller tool is broken and the error is detected in the place p_{10}, then a recovery Petri net block is added to the controller (Fig. 8.13). The error processing procedure includes:

p_{18}: M-1 discarding the old tool;
p_{19}: New tool available; and
p_{20}: M-1 picks up the tool.

There are three new transitions t_{10}, t_{11}, and t_{12}. The error information function $\varepsilon(p_{10}) = 3$.

It is evident that the new controller (including the recovery block) is safe, live, reversible, and essentially decision free (EDF). Given this is true and the initial net is safe, live, reversible, and EDF, then our theorems guarantee that the augmented net has the same properties, i.e., safeness, liveness, reversibility, and EDF property.

As an example of alternate path error recovery, assume that an out-of-tolerance condition for a nut is built up and detected in the place p_{16}. The

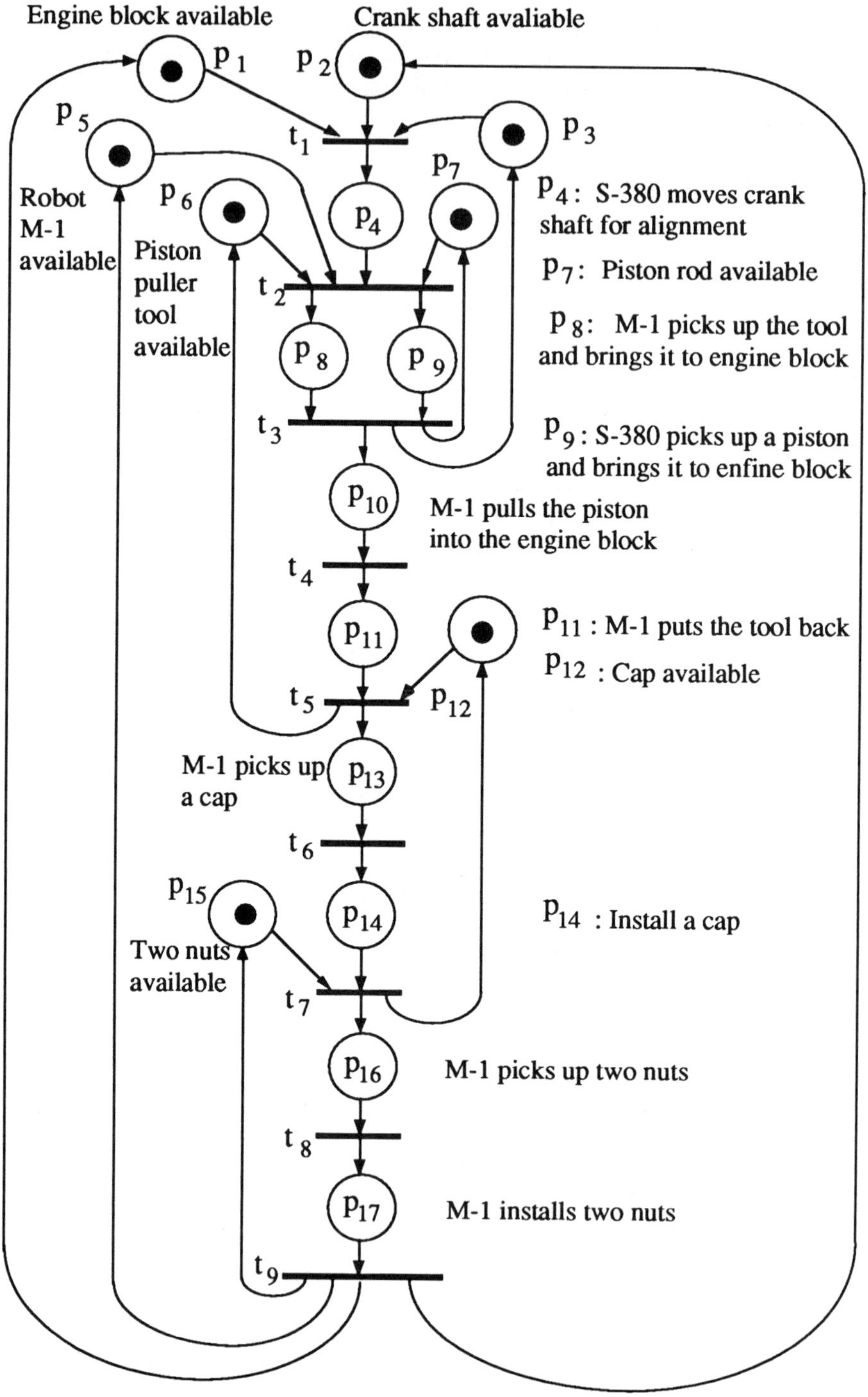

Figure 8.12: A safe, live, and reversible Petri net controller

Table 8.1: Interpretation of places and transitions

Place/Transition	*Interpretation*
p_1	Availability of engine block
p_2	Availability of crank shaft
p_3	Availability of Robot S-380
p_4	S-380 moves crank shaft for alignment
p_5	Availability of Robot M-1
p_6	Availability of piston puller tool
p_7	Availability of piston rod
p_8	M-1 picks up the tool and brings it to engine block
p_9	S-380 picks up a piston and brings it to engine block
p_{10}	M-1 pulls the piston into the engine block
p_{11}	M-1 puts the tool back
p_{12}	Availability of a cap
p_{13}	M-1 picks up a cap
p_{14}	M-1 Installs a cap
p_{15}	Availability of two nuts
p_{16}	M-1 picks up two nuts
p_{17}	M-1 installs two nuts
t_1	Start
t_2 to t_8	Fire on the completion of some appropriate action(s) and when the corresponding resources for the next action(s) are available
t_9	completion of the action in p_{17}

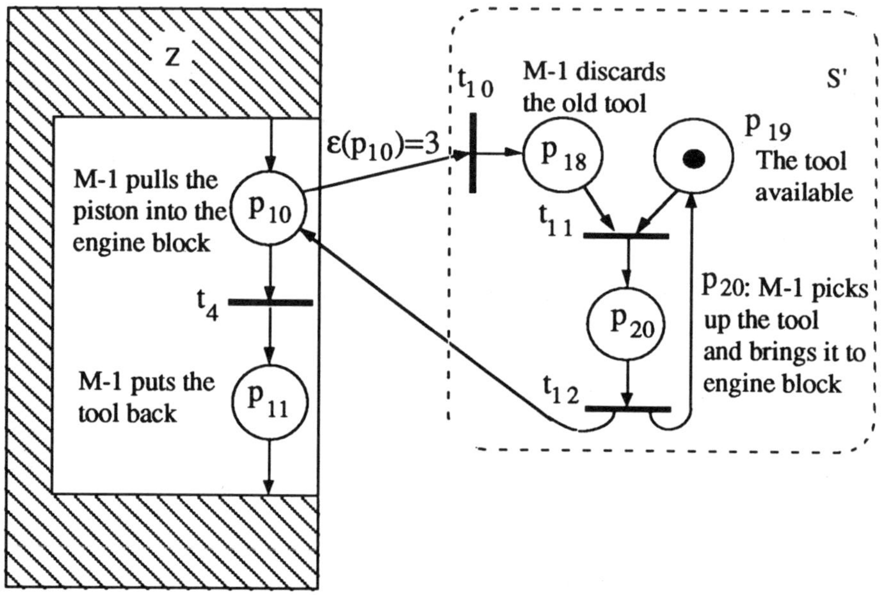

Figure 8.13: Petri net controller after the piston-puller-tool-down error is processed

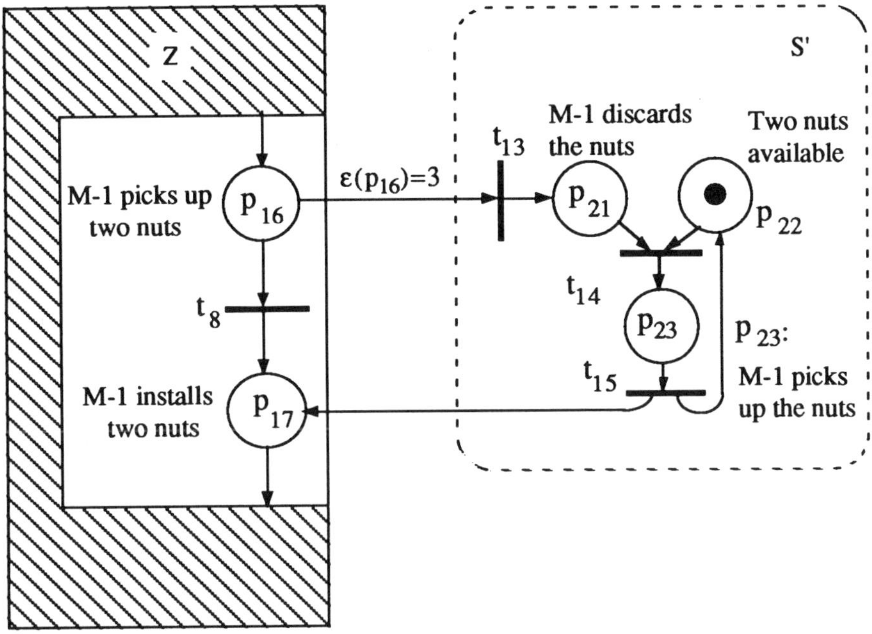

Figure 8.14: Petri net controller after the out-of-tolerance error is processed

error procedure has been represented in Fig. 8.14 which consists of three transitions t_{13}, t_{14}, and t_{15} and three places:

p_{21}: M-1 discards the nuts;
p_{22}: Two nuts available; and
p_{23}: M-1 picks up the nuts.

The error information function $\varepsilon(p_{16}) = 3$ when this error is first-occurred in this place. Again, the liveness, safeness, reversibility, and EDF properties remain unchanged for the controller.

Last, an incorrect part orientation error for a cap is detected in place p_{13}, the error processing procedure can be developed using the forward error recovery method as shown in Fig. 8.15. First the normal state is assumed at the place p_{13}, then the faulty state will be detected in the place p_{14}. The Petri net block which consists of place p_{24}, the robot M-1's adjusting the orientation of the cap, and two transitions, t_{16} and t_{17} is added to the original Petri net controller. We have the error information function

$$\varepsilon(p_{13}) = 3 \text{ and } \varepsilon(p_{14}) = 3.$$

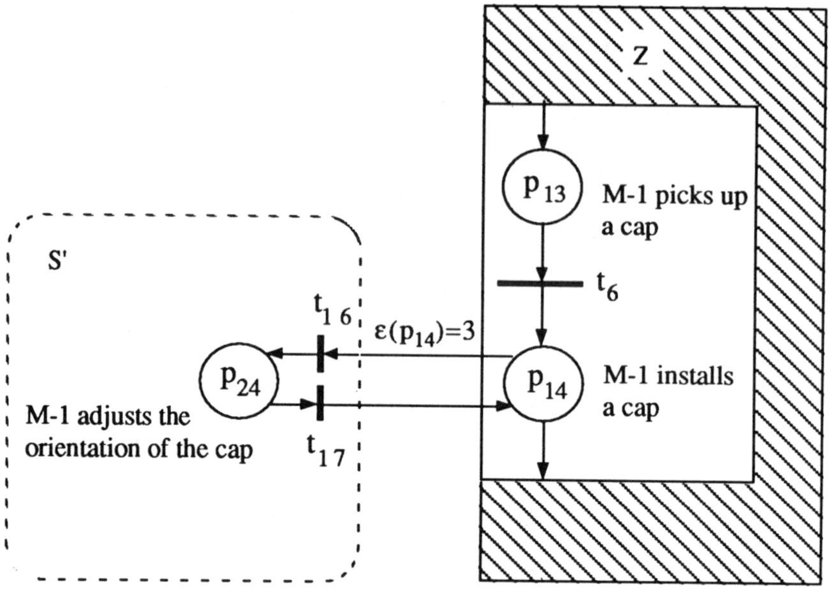

Figure 8.15: Petri net controller after the incorrect part orientation error is processed

Such an augmentation also preserves the properties of safeness, liveness, reversibility, and EDF of the controller.

It is also noted that according to the definition of the error information function, $\varepsilon(q) = 2$ means that emergency processing is needed for any place q and $\varepsilon(q) = 1$ represents that the normal operations are executed.

This section has shown that the errors can be recovered using a Petri net controller. This controller can be adaptively designed using basic augmentation methods, input conditioning method, alternate path method, feedback error recovery method, and forward error recovery method. The desirable properties of the controller in its initial design are preserved as guaranteed by the theorems developed in Section 8.3.5.

8.5 Summary

This chapter has:

1. Presented the concept of augmenting Petri net controllers for error recovery in automated manufacturing systems.

2. Shown that the liveness, boundedness, reversibility and the essentially decision-free property of the nets are preserved using particular program augmentation approaches described by Fielding *et al.* [1988]. They are the input conditioning method, the alternate path method, the backward error recovery method, and the forward error recovery method. Petri nets are successfully used to represent these techniques; thus a formal study can be conducted for analysis of system properties.

3. Presented an example which describe a piston stuffing workstation consisting of two robots and several parts developed at General Motors Technical Center. The following common errors, tools broken or missing, out-of-tolerance parts, incorrect part orientation have been discussed.

The greatest implication of the results is that the behavioral properties of a controller can be maintained when the control code is changed to accommodate new errors. This precludes the need for behavioral analysis each time the code is augmented.

This methodology works when errors can be recovered via local information and dedicated resources. In other word, when global information is needed and shared resources are involved, the error recovery will be more difficult and our methods need to be further extended to those cases.

Chapter 9

Petri Nets: Past, Present and Future

Numerous papers, reports, and books on Petri nets and their applications in various areas have been published since Petri nets were first conceived in 1962. This chapter first attempts to give an overview of Petri net theory development and to discuss briefly some important landmarks from its origin to the present. Next, previous and current applications of Petri nets in manufacturing and automation are discussed. They include modeling, validation, performance analysis, simulation and rapid prototyping of automated manufacturing systems. The use of Petri nets as a tool for designing supervisory controllers for various manufacturing systems, as well as their implementation, is stressed. Research on the synthesis of Petri nets for asynchronous and concurrent systems is summarized in detail. Synthesis methods for deadlock-free control structures are particularly emphasized. Finally, future research in the area of Petri nets for manufacturing automation is discussed.

9.1 Petri Net Theory Development

Net theory was originated by Carl A. Petri in his 1962 doctoral dissertation, where he developed a net approach for modeling and analyzing communication systems. His work became the foundation of a theory of asynchronous information flow and of asynchronous machine models. However, until the late 60s when Holt *et al.* began their Information Systems Theory Project at MIT, net theory did not receive much attention. MIT's limited early involvement had a positive influence on the later rapid development of Petri net theory and its applications. They studied nets of conditions and asynchronous events [Holt 70; Patil 70], and introduced free choice nets [Hack 72]. Petri net languages are defined as sequences of transition firings where each transition is assigned a word [Hack 75]. Deterministic timing is introduced into nets such that system performance can be derived [Ramchandani 74].

In the early development of nets, the analysis of qualitative properties of systems was of central importance. The original tools for conducting analysis are reachability trees or coverability graphs in which all markings embedded in a Petri net are enumerated and relations among these markings are indicated [Karp 69]. Lautenbach [1975] discusses liveness in Petri nets, and first develops the algebraic approaches in which S-invariants or P-invariants and T-invariant can be obtained and then used for analysis of boundedness and liveness. Commoner [1972] studies deadlocks using Petri nets and investigates liveness for arc-weighted free choice nets and simple nets. Using invariant approaches, the analysis of other structural properties is developed by Sifakis [1978].

As the net becomes complex, algorithms for computing invariants are needed. Martinez and Silva [1982] have developed such algorithms for generalized Petri nets. After colored Petri nets are introduced [Jensen 81a], algorithms to calculate invariants also follow [Jensen 81b; Alla 85; Narahari 86]. Meanwhile, reduction methods are developed for the analysis of structural and qualitative properties [Andre 76; Berthelot 76; Lee 85, 87; Murata 80a; Silva 85].

In order to deal with a complex system with many similar components and varied information, Genrich and Lautenbach [1981] formally introduce individual token nets and propose high-level nets, i.e., predicate/transition nets, which include variables as arc labels. Because the variables in these nets yield difficulties when a calculus of invariants is constructed, Jensen [1981a] formulates colored Petri nets with a variable free net calculus. Both kinds of nets provide a more compact graphic representation of a system which may be too complex to describe using ordinary Petri nets.

Basic Petri nets include no notion of time. Time is introduced into Petri nets for the analysis and verification a real-time DES [Merlin 76b; Ramchandani 74; Coolahan 85]. When random variables are used to specify the time behavior, stochastic Petri nets (SPN) result. Appropriate incorporation of temporal and/or stochastic variables into transitions [Ramchandani 74; Ramamoorthy 80] or places [Sifakis 77] in Petri nets can result in a powerful tool for evaluating various systems for temporal or quantitative properties. Molloy [1982] and Florin and Natkin [1982] have independently established the equivalence between a class of stochastic Petri nets and homogeneous Markov processes such that performance can be derived using Markovian analysis techniques. Different net execution polycies are investigated for SPN with transition delays of generalized distributions [Ajmone Marsan 89].

While C. A. Petri [1977] and Genrich *et al.* [1980] continue to develop general net theory, various extended net models have been proposed for a variety of applications. For example, adaptive colored Petri nets have been proposed for designing flexible manufacturing systems (FMS) [Gentina 87]. Petri nets have been extended to include inhibitor arcs or priority rules so that the extended nets exhibit the same modeling power as Turing machines [Peterson 81].

Holt and Commoner [1970] have investigated safeness and liveness of marked graphs which is one of important classes of Petri nets for modeling concurrent systems. Later on, their expansion and reduction methods for preserving safeness and liveness have been reported [Murata 80a; Jonsonbaugh 81], and the synthesis problem for marked graphs has been proposed and solved [Murata 77, 80b, 89].

During the 80s, several books about Petri net theory were published [Peterson 81; Reisig 85; Silva 85a; Yuan 88]. These publications provide good tutorials on Petri net theory and help create a better environment for exploring Petri nets and their possible applications. They have remarked that a mature period of Petri net theory has already come. More and more applications are emerging. The following section discusses one of these applications: Petri nets in manufacturing and automation. Other publications which explore this aspect of Petri net applications can be found in [Al-Jaar 90b; DiCesare 91; Freedman 91; Silva 90; Zhou 90a].

9.2 Applications in Manufacturing and Automation

Broadly speaking, the use of Petri nets in manufacturing and automation can be traced back to the application of Grafcet for modeling finite state automation processes. Grafcet was inspired directly by Petri nets and became a norm for specifying sequential control in France in 1977. Valette *et al.* [1982a] are the first to consider Petri nets as a more powerful tool for specification, validation, and implementation of logic controllers for production systems. Since then, more and more researchers have been involved in the development of Petri net applications for automated manufacturing systems.

In this section, the modeling and validation issues for non-timed Petri nets are first reviewed. Next, an overview of the use of timed and stochastic Petri nets in modeling systems to derive temporal system properties is presented. Some software tools are also reviewed. As a major objective, discrete event control of manufacturing systems is discussed with emphasis on implementation issues. Simulation and rapid prototyping using Petri nets are discussed.

Petri net synthesis methods will be treated as a separate topic discussed later.

9.2.1 Modeling and Validation

In order to explore the applicability of Petri nets for flexible production systems, Valette *et al.* [1982a] have specified and validated interconnected controllers for a transportation system in a car production system using Petri nets. Their experience show that Petri nets are applicable to this system and indicate that such a Petri net approach can be based on decomposition and structuring. Furthermore, Valette [1987] argues that Petri nets are more convenient than other models for concurrency.

Examining the same car production system used by Valette *et al.*, Alla *et al.* [1985] employ colored Petri nets to model the system. Their colored Petri net model is proved to be deadlock-free and bounded. In the same paper, the possible benefit of using colored Petri nets over ordinary Petri nets is revealed; i.e., their conciseness makes it possible to describe a complex FMS with many similarities.

To fully explore the advantages and disadvantages of ordinary Petri nets and colored Petri nets, Martinez *et al.* [1986] have modeled a transfer line consisting of two machines and three buffers, as well as a transportation system consisting of an automatic guided vehicle (AGV), three loading and one unloading stations, using both nets. They briefly discuss the characteristics of a "good model," including boundedness, liveness, and reversibility. Other specific properties include mutually exclusive places and finiteness of synchronic distance.

To understand behavior such as deadlocks and buffer overflows of an FMS, Narahari and Viswanadham [1985] have modeled two manufacturing systems: a transfer line with three machines and two buffers, and an FMS with three machines and two part-types. Boundedness and liveness are analyzed for both systems using invariant methods, and the significance of boundedness, liveness, and reversibility in manufacturing systems is explored. They also present a systematic bottom-up modeling approach which we review later. Furthermore, they demonstrate the usefulness of colored Petri nets in automated manufacturing systems [Viswanadham 87].

To facilitate Petri net modeling and to allow more flexibility, some extended Petri net models have also been proposed. For example, structured adaptive and structured colored adaptive Petri nets include inhibitors, self-modifying arcs, etc., and their usefulness is illustrated through design of an FMS control system [Gentina 87].

A marked graph is decision-free since each place has only one output transition. For ordinary Petri nets, a place can have many output transitions; as a result, different choices in firing enabled transitions may exist. To resolve such conflict situations, Krogh and Sreenivas [1987] add NOT places in a net model to eliminate ambiguity in firing enabled transitions. The net becomes *essentially decision free*; thus, there is no ambiguity in realtime resource allocation for manufacturing processes.

To explore system behavior and to verify logical correctness of specifications, Wang *et al.* [1989] have modeled Connection Management Services of the Manufacturing Message Specification (MMS) using colored Petri nets. The resulting net model offers a hierarchical mathematical description for the entire MMS protocol and can be used for performance analysis of the protocol.

Using Petri nets, it is also possible to describe a coordination model of intelligent machines [Wang 90a, b]. The coordination structure is developed to describe activities and connections among dispatchers and coordinators, each of which is specified as a Petri net transducer capable of translating languages (task plans). Therefore, some system properties can be analyzed.

Based on Petri net theory, software packages have been developed for automatic validation of operations in FMS [Colom 86, 87; Martinez 87]. Cruette and Gentina [1992] have applied Petri nets for design and validation of operation sequences in FMS and Ezpeleta and Martinez [1992] performed formal specification and validation for production plants using high-level Petri nets.

Other work [Beck 86, Jeng 90; Koh 91; Zhou 89, 91, 92] related to the systematic design of Petri nets is discussed later.

9.2.2 Performance Analysis

As previously mentioned, systems are modeled using timed Petri nets in order to conduct temporal performance analysis, i.e., to determine production rates of systems, resource utilization, and the like. It is also possible to detect a bottleneck in an FMS or to determine optimal buffer size, optimal pallet distribution, etc. as indicated by Dubois and Stecke [1983]. In fact, Dubois and Stecke are the first to apply timed Petri nets to describe, model, and analyze production processes. In their research, deterministic time variables are assumed, and a Petri net-based simulation method is utilized to find minimum cycle time and to identify the bottleneck machine for an FMS with three machines and three part-types in a fixed route.

In the *1985 International Workshop on Timed Petri Nets,* many research projects in the field of timed Petri nets are presented, and various applications are reported for analyzing the performance of computer systems [Holliday 85], communication protocols [Gressier 85], and manufacturing systems [Bruno 85]. More importantly, some significant software packages are introduced, for example, DEEP [Dugan 85], and ESP [Cumani 85]. Chiola [1985] presents a user-friendly software package for analysis of generalized stochastic Petri Nets, whose improved version is called GreatSPN [1987]. GreatSPN is powerful in the sense that it can accept various time variables, inhibitors, and random switches and has simulation capacity.

To evaluate the performance of job-shop systems under deterministic and repetitive functioning of a production process, Hillion and Proth [1989] apply a special class of timed Petri nets called timed event-graphs for an FMS with three machines and three job types. The number of jobs in-process is nearly minimized using integer programming, while the system still works at its maximum productivity.

Viswanadham and Narahari [1988a] have provided an excellent introduction to the use of generalized stochastic Petri nets (GSPN) in analyzing the system performance of automated manufacturing systems. They use a software package they developed to evaluate two representative systems: a manufacturing cell with multiple material handling robots, and an FMS with three machines and two part-types.

All these existing software tools mentioned above can be used to analyze the performance of automated manufacturing systems. For example, Al-Jaar and Desrochers [1990b] investigate the performance of transfer lines and production networks using SPNP developed by Ciardo [1989]. Zhou and Leu [1991b] utilize SPNP to evaluate the performance of a two robotic manipulator assembly station for printed circuit boards.

Aiming at demonstrating applicability of and accuracy achieved by stochastic Petri nets for FMS, Watson and Desrochers [1991] perform three representative case-studies from different sources using SPNP and GreatSPN. The work shows that the performance analysis results obtained using Petri nets agree with those obtained by simulation [Dunkler 88], queuing theory [Yao 85], and probability theory [Seidmann 89].

SPNP continues to be used for analyzing the performance of manufacturing systems which have fixed routing and produce limited types of parts. Zhou *et al.* [1990d] have first developed a stochastic Petri net modeling approach such that the good behavioral characteristics of the model can be preserved. Then they analyze deadlock-free and deadlock-prone manufacturing systems. Their results show that supervisory controllers with freedom from deadlock are better than deadlock-prone controllers in designing real-time resource-sharing distributed systems.

Recent developments in performance analysis of automated manufacturing systems has led researchers to investigating FMS with possible deadlocks [Viswanadham 90]. One reason for this investigation is that it can be very difficult to design a deadlock-free controller for complex systems with numerous job types.

Moment generating functions and Petri nets have been combined to obtain symbolic performance solutions for stochastic Petri nets, generalized stochastic Petri nets, and extended stochastic Petri nets [Guo 91, 92; Zhou 92d].

Reduction methods are used to derive the closed form of the performance indices [DiCesare 92].

Although the underlying models of stochastic Petri nets are still Markov processes, the Petri net approaches preclude the direct construction of the state space required by Markovian analysis. This is often too large for humans to manipulate. In order to deal with a stochastic net of a large state space, reduction and approximation methods have been investigated [Jungnitz 91, 92; Ma 92]. Some reduction techniques for ordinary Petri nets have been extended to the generalized stochastic Petri nets to make it possible to analyze the performance of large-scale discrete event systems. Queuing network methods and GSPN have been combined to model and analyze complex systems [Balbo 87, 88]. A decomposition method has been developed to evaluate SPN models and applied to a manufacturing system [Ciardo 91]. Performance bounds for some classes of timed Petri nets have been derived in polynomial time [Campos 90] and the method is used to evaluate repetitive automated manufacturing systems.

9.2.3 Control of Manufacturing Systems

Around 1980, various researchers independently studied Petri net-based methods to develop a programmable logic controller (PLC) [Bollinger 88]. Silva and Velilla [1982] present a comparative study of various implementations and indicate that one benefit of using Petri net approaches is that both software and hardware errors can be prevented in the PLC design phase because of the possibility of early validation of controllers.

To put such methods to use, Courvoisier *et al.* [1983] develop a prototype system for designing a colored Petri net-based PLC and deriving control code. This system is used for designing a system with two AGVs sharing a common path.

To effectively handle the complexity of FMS control, Valette *et al.* [1985a] propose a hierarchical control structure which comprise local control level, coordination level, and monitoring and real-time shop scheduling level. Communication among different PLCs is attained by a local area network in this type of architecture. Merabe [1986] also proposes and applies a Petri net based approach to synchronization of operations in FMS.

One of successful applications of Petri net approaches to system control was reported by Murata *et al.* [1986]. First, Petri nets are modified into Control nets which are a class of safe Petri nets. Then Control nets are used to design station controllers for sequencing control. The station coordinators for monitoring and diagnosis are implemented using IF-THEN rules. Two objectives - flexibility and maintainability are achieved when such a Petri net approach is used. In fact, their research has produced Hitachi commercial systems.

To accommodate changes in FMS, Menon *et al.* [1988] have developed a system for configuring coordination controllers in FMS using colored Petri nets. Also, they present a pilot Petri net workstation controller.

Crockett and his colleagues [1987] are able to directly implement a Petri net controller for a manufacturing workstation without need of a PLC. In their controller, the token player idea is applied. The controller is application independent for a general purpose computer and uses Petri nets to describe the sequencing information.

To deal with more complicated manufacturing systems, Kasturia *et al.* [1988] have extended Crockett's direct implementation method to colored Petri nets. Their proposed controller is implemented in real-time on a multi-level manufacturing system which consists of two workstations. Jafari [1992] has also proposed an architecture for a shop-floor controller using colored Petri nets with ordered colored sets and structured input and output functions.

Instead of using Petri nets to design control logic, Krogh *et al.* [1988] has developed a software tool which converts specifications into a Petri net to validate their consistency and viability and generates the control source code for discrete manufacturing systems.

The current trend is to combine process plans, scheduling, and dispatching with manufacturing resources to derive a flexible Petri net controller for a more and more complicated manufacturing system. Artificial intelligence techniques and Petri nets can be combined to achieve real-time planning, scheduling and control of FMS [Lee 92; Martinez 88; Zhang 92; Zhou 91d]. Instead of a centralized control structure, an open distributed control architecture is one focus of future research.

9.2.4 Simulation and Rapid Prototyping

Simulation is needed when the system is very complicated and computationally intractable using analytical methods. It is sometimes necessary at the final system design stage. Petri net models of a discrete event system can be easily used to drive discrete event simulation. Valette *et al.* [1985b] develop a colored Petri net simulator using PASCAL. This simulator is able to analyze the system performance by associating the time durations to places.

Graphical visualization is a key advantage with Petri net methods. GreatSPN [Chialo 87] is the first package which includes a convenient graphical user interface for Petri net editing on Sun workstations. In addition to its ability to analyze generalized stochastic Petri nets analytically, it is able to simulate the net to derive the performance of a system with generalized distributions. Similar user friendly and interactive graphical tools for Petri nets have been also developed by Shukla and Robbi [1991].

Bruno and Morisio [1986] are the first to take the advantage of object-oriented programming language to implement a Petri net simulator for the rapid prototyping of process control systems. They utilize PROT (Process-Translatable) nets which is an extension of Petri nets to achieve the purpose. The functions include defining objects, editing nets in graphics, compiling nets into executable files, and simulating the models with animation.

Simulation language is able to represent various complicated execution rules for Petri nets. Beck and Krogh [1986] have successfully utilized SIMAN to simulate their modified Petri net models for manufacturing systems.

By recognizing the need in manufacturing system design and the popularity of personal computers, researchers [Chen 91; Tang 92] have developed a graphic based simulator on PCs. Their developed package has the ability to predicate the performance of deterministic time Petri nets and to simulate real-time control logic for flexible manufacturing systems.

Other work includes the application of Petri net based simulation to detect conflicts for AGV routing in FMS [Zeng 91].

9.2.5 Petri Nets for Error Recovery

Fault detection and error recovery play a paramount role in real-world systems such as automated manufacturing and communication. Merlin and Farber [1976a] are the first to model a recoverability problem in a communication protocol using timed Petri nets. Errors like token-loss can be recovered by designing a system properly although such a design depends on a priori knowledge of the execution time of the events in the system.

To implement fault-tolerant systems, Sifakis [1979] has applied error detection and correcting codes to Petri nets. Silva and Velilla [1985b] have extended his work using linear codes and adding implicit and test places in Petri nets. Furthermore, they propose a generalized watch-dog and implement it for error detection of an industrial manufacturing system using Petri nets [Velilla 88].

Morganti [1982] utilizes Petri nets in the design and implementation of error recovery strategies, and in the analysis and evaluation of fault-tolerant properties of redundant design. At the same time, Valette [1982b] suggests Petri nets for modeling the reliability of real-time systems.

Viswanadham and Johnson [1988b] provide an error diagnosis scheme based on Petri nets and fault trees for automated manufacturing systems. In their scheme, there are dedicated diagnostics for subsystems and system diagnostics for the system level. The system level is implemented by a Petri net controller, which monitors part flows and coordinates the local diagnostic systems.

Sahraoui *et al.* [1987] have studied the similarities between a Petri net and a knowledge-based system and proposed a method which combines Petri nets and artificial intelligence techniques for error detection and diagnosis. Their research emphasis is on the effective monitoring of a flexible manufacturing system. Some other machine intelligence approaches to automatic error recovery have also been reported [Srinivas 78; Kumaradjaja 89; Chang 90, 91]. A review of these approaches is given by Fielding *et al.* [1988, 1989].

9.3 Petri Net Design Methods for DES

This section reviews various previous efforts to systematically model discrete event systems using Petri nets. The emphasis is put on the motivations, achievements, and limitations of researchers' approaches in this area. A review of these previous methods is also given by Jeng and DiCesare [1990], where the technical aspects are stressed and several manufacturing examples are presented.

Previous research on Petri net design methods falls into two basic categories: bottom-up and top-down. Bottom-up approaches begin with the construction of subnets for component processes, and proceed to the final net by merging and/or linking all these subnets.

Top-down synthesis is characterized by the stepwise refinement of an aggregate Petri net model. Each successive refinement contains increasing detail until the implementation level is reached. Since reduction methods are closely related to top-down synthesis, they are also reviewed briefly in this section.

9.3.1 Bottom-up Approaches

Since their invention by Lautenbach in 1972, invariant methods have been proved to be powerful in analyzing such qualitative properties as boundedness and liveness. However, the construction of all invariants for a large net is costly in computation. Assuming that a global net results from a combination of its subnets, an alternative is to compute first the invariants of all subnets, and then the invariants of the global net. This method becomes promising if the invariants of the global net can be easily computed using those subnets' invariants. Agerwala and Choed-Amphai [1978] are the first to notice possible solution for synthesizing a concurrent system. They propose a set of synthesis rules for the bottom-up construction of a complete Petri net from subnets which share common places. These rules provide an iterative algorithm for calculating the resulting net's invariants, which can be used to analyze some Petri net properties.

Along a similar line, Narahari and Viswanadham [1985] have studied the synthesis of a Petri net model of a flexible manufacturing system (FMS) and presented a systematic bottom-up approach. They obtain their Petri net model by constructing a sub-Petri net model for each machine operation and then combining these subnets by sharing places. It is proven that the p-invariants of the resulting Petri net are the union of the p-invariants of the subnets under their proposed conditions.

These studies make it easier to compute the invariants of some types of large nets. These invariants can be used to derive boundedness and liveness of nets given an initial marking. However, other properties such as reversibility remain difficult to analyze since the invariants of a Petri net do not convey complete information about the structure of the net [Agerwala 78]. Moreover, analysis using invariant methods becomes numerically more difficult as a Petri net becomes larger, since equations for invariants become more numerous. For example, a Petri net model for a two-workstation manufacturing system is analyzed by Kasturia *et al.* [1988] using the invariant method. The incidence matrix has 58 rows and 42 columns. Forty equations are obtained for the p-invariants and sixty-one for the t-invariants.

Noticing the possible cumbersome analysis that results from using invariants, Krogh and Beck [1986] have adopted another method and made a significant contribution to bottom-up synthesis in the sense that (a) their bottom-up approach can guarantee safeness and liveness of the resulting net, and (b) they extend common place or transition sharing concepts to common path sharing. In other words, their method can avoid using invariants for the verification of safeness and liveness of a Petri net. They introduce Modified Petri nets (MPNs) first, then decompose a manufacturing process into operations and resources, and model the activities of each resource as an elementary MPN circuit. A safe and live Petri net model results from the union of elementary MPN circuits along common paths.

Similar work is done by Ahuja and Valavanis [Ahuja 88; Valavanis 90], who introduce their extended Petri nets in which each of tokens, places, transitions, and arcs differs from the other so that more information can be carried in a net model of the same structure.

However, the assumption that each resource's activities can be modeled by an elementary circuit restricts the application of this approach. The approach cannot deal with the situation where bounded, but not safe places exist. However, these situations may be common in a manufacturing system, for example, when the system includes a buffer with a capacity greater than one. It is also difficult to deal with resource-sharing cases using this method.

Along the same line, Koh and DiCesare [1990, 1991a, b] introduce the live and bounded circuit concept and formulate three transformations to synthesize live and bounded Petri nets. Using their results, a generalized Petri net having liveness and boundedness can be synthesized by sharing common paths. The method is also applicable to Petri net reduction. However, the method does not include general resource-sharing situations.

In order to design complex logic controllers, an incremental approach [Ferrarini 92] has been proposed by using condition/event Petri nets as their building blocks. Single control tasks are implemented and specified with Petri nets first, then appropriate connections are established to coordinate and synchronize these tasks. The theory developed can guarantee desired properties of the overall system under several commonly-used connections. For the connections which result in closed paths, the further investigation is needed.

Recently, Jeng and DiCesare [1992a, b] have developed a modular method which is able to identify resource places whose initial number of tokens influences the properties of the net. An algorithm has been devised to reach a conclusion whether the net has an undesirable structure such as a circular resource waiting circuit. Then remedies can be taken to avoid deadlock for many systems.

Other related work is reported by Villarroel *et al.* [1988]. A graphic system for manufacturing design is developed, in which a colored Petri net can be generated from the descriptions of a plant and a process plan.

Summarizing, there is no bottom-up synthesis procedure which guarantees the properties of liveness, boundedness, and reversibility for the general ordinary Petri net case [Zhou 90a].

9.3.2 Top-down Approaches

Before previous top-down synthesis methods are reviewed, reduction methods are briefly discussed because:

1. Some reduction rules may be inversely used to expand a Petri net, and this expansion often maintains desirable properties such as boundedness and liveness; and

2. Reduction methods unveil the significance of a top-down design strategy, in which a Petri net grows each refinement step which preserves its qualitative properties.

Petri net reduction is a procedure for transforming Petri nets to reduced nets, preserving some of the desirable properties of the original. Early work on Petri net reduction is done by Hack [1972]. Recognizing the challenge of analyzing a complicated Petri net using reachability trees, Andre *et al.* [1976] have formulated a reduction method by replacing a subnet of a state machine with a macroplace. The method is modified and extended to generalized Petri nets by Silva [1981, 1985].

Many reduction methods are proposed to prove the properties of parallel programs in the literature. Berthelot [1976] has proposed a more powerful reduction procedure, which is realized by cutting arcs and replacing subnets with places and transitions. Furthermore, Berthelot [1986, 1987] formulates a set of transformations to simplify or refine a Petri net without changing important properties; and he also provides a decomposition technique for splitting a system into subsystems which can be analyzed separately.

To be able to reconstruct any intermediate net in a reduction process, Lee and Favrel [1985] provide a reduction method which allows a hierarchical decomposition of a large system into multilevel subsystems while the liveness, boundedness, and proper termination of the net are maintained. They have subsequently extended the results to generalized Petri nets [1987].

Most of these methods for the reduction of Petri nets can also be applied to the synthesis of large Petri nets. Using the step-by-step refinement idea, Valette [1979] presents the conditions under which a subnet can be substituted for a single transition while preserving properties such as liveness and boundedness. Suzuki and Murata [1980, 1983a] generalize the results obtained by Valette. Their results can work for two directions: reduction or abstraction, and expansion or synthesis of Petri nets. Zhou *et al.* [1989b] have found that the stepwise refinement is particularly appropriate for the top-down design of Petri net models and controllers for real-life systems.

Murata [1977] is the first to propose the synthesis problem for a class of Petri nets called marked graphs [Commoner 71], in which each place has exactly one input transition and one output transition. Using parallel, series or sequence, and circuit reductions and expansions on a marked graph, such properties as safeness, liveness, and reversibility of the graph can be proved to be preserved [Murata 80a], and three additional such transformations are provided [Jonsonbaugh 81]. Beginning with a safe and live marked graph, their transformations can generate a safe and live marked graph. Furthermore, Murata [1980b] proposes a modular marked graph synthesis method for decision-free concurrent systems. The use of this approach ensures that the properties of marked graph models such as liveness, safeness, the maximum resource requirement, and computation rate are prescribed.

Berthelot [1986] has proved that live and bounded marked graphs or live, bounded and persistant Petri nets can be transformed into a net of a single transition using his proposed transformations. This implies that live and bounded marked graphs can be synthesized beginning with a single transition.

Datta and Ghosh [1984b] also present a modular approach to synthesizing bounded and live Petri nets by selecting regular nets as their basic modules and interconnecting the individual modules following their proposed rules. Their results are also used to reduce large Petri net models of communication protocols [1986].

Chao *et al.* [1992] formulate a knitting technique for Petri net synthesis by adopting a global and incremental synthesis approach. The idea behind this technique is to introduce details in an incremental way, i.e., by adding new paths and/or cycles to a Petri net. At each step the Petri net model grows according to certain rules which will guarantee such system properties as boundedness, liveness, and reversibility. Thus the cumbersome analysis for these properties can be avoided while designers can still build up a Petri net model for a complicated system.

Early work in synthesis approaches is also reported by Yoeli [1980]. A general synthesis procedure of Petri nets was discussed based on top-down and decomposition ideas starting from system specifications.

The above discussion serves to point out that a design problem is more challenging than a reduction problem in the sense that a Petri net structure and initial markings need to be constructed from the given specification of a system, while the net structure and initial markings are generally given for a reduction problem. Another reason why a design problem is more challenging is that researchers may not necessarily care about the physical significance of each step in a reduction.

9.4 Petri Nets in the Future

This book presents recent developments in Petri net synthesis methods, discrete event control, automatic error recovery, and their applications to automated manufacturing systems. It has demonstrated that Petri nets are a very useful tool for discrete event control of manufacturing systems. To assure their successful applications to industrial automation and achieve their wide acceptance by application engineers, work must be done in education and computer aided design (CAD) tool development. Education in the area of Petri nets or more generally discrete event systems, plays a very important role in equipping engineers with modern knowledge and tools in the design of advanced automated systems. Numerous universities and colleges have begun to offer basic and advanced courses in the areas of Petri nets and discrete event systems. Experts have given the tutorials and workshops in various international conferences and symposia.

User-friendly CAD tools will ease the usage of sophisticated theory for many industrial applications in manufacturing systems design, communication network design, and office automation. As we see in this book, the human validation of some conditions in our theory may become very difficult. A computer-based automatic verification procedure needs to be developed such that designers can be relieved of tedious validation work. A Petri net tool including the following important components as shown in Figure 9.1 is very

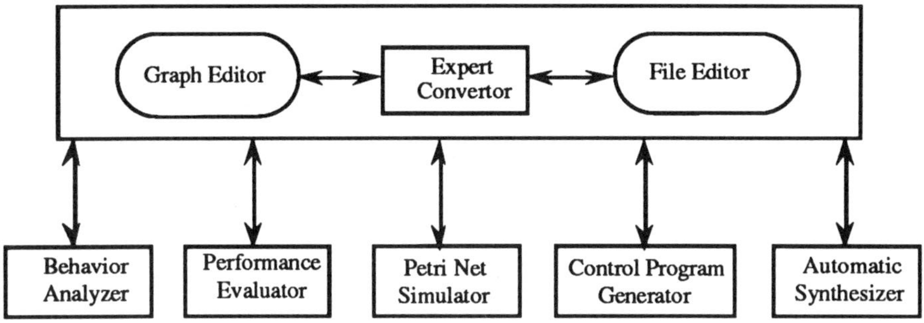

Figure 9.1: A CAD tool for manufacturing system design using Petri nets

desirable [Zhou 89c, 92e]:

1. A powerful and user-friendly Petri net graphic editor

2. A behavior analyzer via reachability graph generation, invariant methods, or reduction

3. A performance evaluator via Markov processes, transfer functions and simulation

4. A simulator of Petri nets with dynamic graphical display

5. A supervisory control program generator/interpreter

6. An automatic Petri net synthesizer.

The major tasks in the development of each component in such an integrated tool are outlined as follows.

Petri Net Graphic Editor

A well-designed graphical user interface (GUI) is the key to successful application of any software tool. Algorithms for curved arcs including the generation of B-spline, conic-spline, and special feature curves need to be developed. One design objective is to minimize the number of arc intersections. After a user designs a Petri net, a graph optimizer or drawing expert can be used to achieve a "pretty" Petri net based on aesthetic principles.

Behavior Analyzer

Important aspects of manufacturing system design are the analysis of its qualitative properties and the verification of its logical correctness. One advantage of applying Petri net to the study of manufacturing systems is their mathematical capacity to check a modeled system's structural properties. Verification of these properties is implemented by four methods: reachability generation, invariant method, simulation, and reduction algorithms. When the numbers of places, transitions or tokens are large, the problem of state explosion excludes the reachability graph approach; thus either simulation or reduction approaches must be used. The related algorithms need to be developed or implemented.

Performance Evaluator

Temporal performance, like throughput and machine utilization, is used in evaluating manufacturing systems. If time delays are exponentially distributed random variables, performance can be efficiently evaluated by numerically solving an underlying Markov chain (assuming a tractable number of states). Otherwise, simulation or approximation is needed. In order to obtain symbolic performance solutions, a transfer function-based approach [Guo 92] has been developed and is being implemented in software [Akella 92]. This method will be extended to generalized distributions.

Petri Net Simulator

A Petri net simulator simulates execution of a Petri net, i.e., the flow of tokens in the places of the net through transitions. Simulation gives a vivid graphic description of a system's operation to aid in model design and debugging. Simulation becomes necessary when the performance cannot be predicted by the system performance evaluator described.

Supervisory Control Program Generator

One of the keys to industrial application of a tool for design of manufacturing systems is to automatically generate the control code given a Petri net definition. One central task is to develop a Petri net compiler for Petri net controllers such that a Petri net representation can be converted into a Programmable-Logic-Controller-like supervisor for industrial applications. The difficulty lies in the fact that building blocks including hardware components have not been standardized. Therefore, to develop a compiler for general purposes becomes a challenging problem. A possible architecture includes a Petri net compiler, a Petri net interpreter, a Petri net on-line monitor, database, and knowledge base [Zhou 92c]. In addition to the compiler idea, this architecture allows the use of a Petri net interpreter originating from the token player concept. Each time a decision is made to fire an enabled transition, a specific operation in a manufacturing system is triggered as Chapter 7 shows.

Automatic Synthesizer

Given discrete event system specifications, the automatic generation of Petri net models using existing synthesis concepts and algorithms is desirable. The synthesizer will be able to guide a user to construct a correct model while checking for semantic and syntactical errors.

To meet with increasing needs of event-driven systems for automation in today's world and to fully use the theory developed in the area of Petri nets, such a CAD tool has to be developed. It promises to help achieve flexible, efficient, high-quality and low-cost production for many industrial applications. Furthermore, students and engineers are able to learn the theory and knowledge of Petri nets by using such a tool.

The theoretical investigations are needed in the following areas:

- Feasible initial marking problem,
- Multi-layer resource-sharing: incorporation with process planning and scheduling,
- Extension to high-level Petri nets, and
- Relations to other discrete event system models.

The Feasible Initial Marking Problem

A very important theoretical problem is to find all feasible initial markings given a Petri net structure with a classification of places such that the marked Petri net is bounded, live, and reversible. This topic has evident implications in designing Petri nets for a distributed system. For example, the solution can be used to answer how many resources are allowed for a particular type of resources such that the system is deadlock-free and cyclic if all the other types are fixed.

Multi-layer Resource-sharing: Incorporation with Process Planning and Scheduling

A flexible manufacturing system consists of many machines and can deal with many types of jobs. The job processes share all machines. Furthermore, when designers need to build a real-time FMS, they need to consider transportation systems, tool-sharing, robot-sharing, and so on. Obviously these components are shared by the machines. Therefore, two layers of resource-sharing are formed. The first layer is concerned with the process planning and scheduling where machine intelligence techniques are often used to reach a good policy. The second layer focuses on a physical implementation of the planning results given a list of resources, i.e., machines, shared transportation systems, and robots.

Designers can specify the entire system as a Petri net model which can be "compiled" into a controller to achieve the system performance and properties. However, modeling of a Petri net with desirable properties becomes extremely difficult. More research work needs to be done in this area.

Extension to High-level Petri Nets

In high-level Petri nets such as colored Petri nets, many similar structures can be represented as an identical structure with different tokens in a place. They can make net representation more compact than ordinary Petri nets. The problem is that an appropriate trade-off between compactness and understanding needs to be made. This is because the more compact a net, the more difficult it is to visualize its properties.

Relations to Other Discrete Event System Models

Other methods, such as communicating sequential processes [Hoare 85], finitely recursive processes [Inan 88], formal language and automata based approaches [Wonham 84, 87; Ramadge 83, 89], queueing networks [Yao 85], and perturbation analysis techniques [Ho 79, 87] need to be further investigated for modeling and control purposes. In order to fully use the advantages of these models for designing discrete event systems, researchers need to understand the underlying relationships among these models and Petri nets [Giua 91].

Appendix A
Proof of Lemma 4.4

Lemma 4.4 *If C0 is true, then Z' is live with respect to $m'_0 = (m_0, 1)$ if Z is live with respect to m_0.*

Proof:

Let Z^k denote the Petri net with an SME $(p_E, D^{(k)}), D^{(k)} = D^1 \cup D^2 \cup ... \cup D^k$ for $k = 1, 2, ..., n+1$. $T_a = T_a^1 \cup T_a^2 \cup ... \cup T_a^{n+1}$, where $T_a^j = \{t_{aj}^i, j \in N_{ki}\}$. Z'_i results from Z' with $I(p_E, t) = O(p_E, t) = 0, \forall t \in D - D^i$. $m[f>$, $m[g>$, etc. represent a marking resulting from the firing of f at m. The proof is conducted by induction on k as follows.

First, when $k = 1$, this case deals with the addition of k_1-PME. The conclusions hold true using Theorem 3.2.

Second, suppose that $k \leq n$, the conclusion is true. The following shows that it is so for $k = n + 1$.

$\forall m \in R(Z^{n+1}, m'_0)$ and $t \in T$, we need to show that $\exists f$ in $Z^{n+1}, \ni m[f>$ enables t. Since $m \in R(Z^{n+1}, m'_0), \exists g, \ni m'_0[g> m = (\mu, m(p_E))$. Since either m or $(\mu, 1) \in R(Z^n, m'_0)$ and since Z^n is live according to the induction assumption, $\exists f$ in Z^n, $\ni m[f>$ or $(\mu, 1)[f>$ enables t where each transition is necessary to enable t. Concentrating on the first case, the following proof can be extended to the other case.

There are two subcases:

A. f is fireable in Z^{n+1},

B. f is not fireable in Z^{n+1}.

A. f is fireable in Z^{n+1}.

A.1: If $m[f>$ enables t in Z^{n+1}, we are done.
A.2: If $m[f>$ does not enable t in Z^{n+1}, then noticing the difference between Z^n and Z^{n+1}, we have the following possible case only:

$$m(p_E)[f> = 0 \text{ and } t \in T_a^{n+1}.$$

This implies that $\exists(t_a^{n+1}, t_b^{n+1}) \in D^{n+1}, \ni \sharp(f, t_a^{n+1}) = \sharp(f, t_b^{n+1}) + 1$, and the last transition from T_a in f is t_a^{n+1} since f is fireable in Z^{n+1}. Note that Strictly, $\sharp(gf, t_a^{n+1}) = \sharp(gf, t_b^{n+1}) + 1$ where $m_0[g> m$. For simplicity, we just consider $\sharp(f, t_a^i) = \sharp(f, t_b^i) + 1$.

A.2-1: If $\exists h$ in Z^n that is fireable in Z^{n+1}, and $m[fht_b^{n+1}> m'$ enables t, we are done.

A.2-2: Otherwise, $m'(p_E) = 1$ and t is disabled. For such $m', \exists h', \ni t$ is enabled before its firing in Z^n. Such *a* g is fireable in Z^{n+1}. If $m'[h'> m''$ with $m''(p_E) = 0$, then $\sharp(f, t_a^{n+1}) = \sharp(f, t_b^{n+1}) + 1$ where $(t_a^i, t_b^i) \in D^i$. For such m'', using Lemma 4.3, t_b^i can be enabled and fire in Z^{n+1}. Suppose that the resulting marking is m^*. Thus, at m^*, t is enabled.

B. f is not fireable in Z^{n+1}.

The difference between Z^n and Z^{n+1} makes the first disabled transition of f in Z^{n+1} be a transition in T_a, say t'. Let $f = f_1 t' f_2$. Then $\exists(t_a, t_b) \in D^{(n+1)}, \ni \sharp(f_1, t_a) = \sharp(f_1, t_b) + 1$ and t_a is the last transition from T_a in f_1. Also, at least one of t' and t_a is from T_a^{n+1}. If the number of transitions in T_a which are not fireable in f is K, our objective is to find a new sequence such that it is fireable and enables t in Z^n, but has $K - 1$ unfireable transitions from T_a. Three cases need to be considered: $EP(t_a, t')$ is an A-path, $EP(t', t_a)$ is an A-path, and neither $EP(t_a, t')$ nor $EP(t', t_a)$ is an A-path.

B.1: $EP(t_a, t')$ is an A-path. This case guarantees $t' \in T_a^{n+1}$.

B.1-1: If f_2 contains t_b, f_2 can be arranged as $f_1 f_{21} t_b t' f_{22}$ where f_{21} is the necessary transitions in f_2 to fire t_b, and f_{22} is f_2 with the removal of t_b

and f_{21}. According to the definition of an SME, firing t_b does not need the firing of t'. Also, $t_b's$ firing does not disable t'. $f_1 f_{21} t_b t'$ is fireable in both Z^n and Z^{n+1}. Also, $m[f_1 f_{21} t_b t' f_{22}>$ enables t in Z^n.

Similarly, if f_2 contains t'_b where $(t', t'_b) \in D^{n+1}$, the required sequence can be found.

B.1-2: If f_2 contains neither t_b nor t'_b, since Z^n is live, there exist h and h' which are necessary to enable t_b and t'_b from $m[f>$, respectively. It can be easily proved that neither h nor h' contains any transition in T_a. If $m[f_1 t' f_2 h t_b>$ or $m[f_1 t' f_2 h t'_b>$ enables t in Z^n, we are done by using B.1-1. Otherwise, if both disable t, we will show a contradiction. There are three cases.

a. $t \in EP(t_a, t_b)$. This implies that an A-path $EP(t_a, t)$ exists. The firing of t is necessary to fire t_b. Two cases follow. The first is that if no A-path $EP(t_b, t')$ exists, then $\exists p \in P_C, EP(t', t), EP(t, t_b)$, and $EP(t_b, p)$ make $EP(t', p)$ which has no t'_b in Z'_{n+1}. This is a contradiction. The second is that if there is an A-path $EP(t_b, t')$, then from $m_0[g f_1 t' f_2>, \sharp(g f_1, t_b) - \sharp(g f_1, t') \leq C(t_b, t') - 1$ since f_1 contains t_a. Since $t_b's$ firing needs $t's$ firing at $m_0[g f_1>$ and $t's$ firing needs t', the token capacity between t_b and t' would be achievable. This is a contradiction.

b. $t \in EP(t', t'_b)$. Similar to the above reasoning, we can obtain contradictions.

c. $t \notin EP(t_a, t_b) \cup EP(t', t'_b)$. Since $m[f_1 t' f_2 h t_b>$ or $m[f_1 t' f_2 h t'_b>$ disables t in Z^n, $\exists p, p' \in P', \ni EP(p, tb) \neq \emptyset, EP(p, t) \neq \emptyset$, and $EP(p', t'_b) \neq \emptyset, EP(p', t) \neq \emptyset$. If either $EP(t, p) \neq \emptyset$ or $EP(t, p') \neq \emptyset$, then the existence of $EP(t_a, t)$ and $EP(t', t)$ makes $t \in EP(t_a, t_b) \cup EP(t', t'_b)$. This is a contradiction. Otherwise if $EP(t, p) \neq \emptyset$ and $EP(t, p') \neq \emptyset$, then, in Z'_{n+1}, after the firing of t, t'_b cannot fire since no token is in p'. This is contrary to that D^{n+1} forms a PME in Z'_{n+1}.

B.2: $EP(t', t_a)$ is an A-path.

Similar to the proof for B.1.

B.3: Neither $EP(t_a, t')$ nor $EP(t', t_a)$ is an A-path.

One case is that $t_a, t' \in D^{n+1}$. The other case is that two transitions are from different PMEs and has no A-path between them. Both can be proved using the method similar to B.1.

Since each time we reduce one unfireable transition in T_a from the original sequence to the new one through the above procedure, we finally have a fireable sequence in Z^{n+1} that enables t in Z^n. This is the case A.
QED

Bibliography

[Agerwala 78] Agerwala, T. and Y. Choed-Amphai (1978). "A synthesis rule for concurrent systems," in *Proc. of 15th Design Automation Conference,* Las Vegas, pp. 305-311.

[Agerwala 79] Agerwala, T. (1979). "Putting Petri nets to work," *Computer*, 12(12), pp. 85-94.

[Ahuja 88] Ahuja, J. S. and K. P. Valavanis (1988). "A hierarchical modeling methodology for flexible manufacturing systems using extended Petri nets," in *Proc. Rensselaer's 1st Int. Conf. on Computer Integrated Manufacturing*, pp. 350-356, Troy, NY.

[Al-Jaar 89] Al-Jaar, R. Y. (1989. *Performance Evaluation of Automated Manufacturing Systems Using Generalized Stochastic Petri Nets*, Doctoral Dissertation, ECSE, Rensselaer Polytechnic Institute, Troy, NY.

[Al-Jaar 90a] Al-Jaar,R. Y. and A. A.Desrochers (1990a). "Petri nets in automation and manufacturing," in G. N. Saridis (ed.), *Advances in Automation and Robotics*, 2, JAI Press Inc., pp. 153-225.

[Al-Jaar 90b] Al-Jaar, R. Y. and A. A. Desrochers (1990b). "Performance evaluation of automated manufacturing systems using generalized stochastic Petri nets," *IEEE Trans. on Robotics and Automation*, 6(6), pp. 621-639.

[Ajmone Marsan 89] Ajmone Marsan, A., G. Balbo, A. Bobbio, G. Chiola, G. Conte, and A. Cumani (1989). "The effects of execution policies on the semantics and analysis of stochastic Petri nets," *IEEE Trans. on Software Engineering*, 15(7), pp. 832-846.

[Akella 92] Akella, P. and F. DiCesare (1992). "Closed form analysis with Petri nets," in *Proc. 1992 IEEE Int. Conf. on Systems, Man, and Cybernetics*, Chicago, IL, pp. 644-649.

[Alla 85] Alla, H., P. Ladet, J. Martinez, and M. Silva (1985). "Modeling and validation of complex systems by colored Petri nets: application to a flexible manufacturing system," in *Advances in Petri Nets 1984*, G. Rozenberg, H. Genrich, and G. Roucairol (ed.), Springer-Verlag, pp. 15-31.

[Andre 76] Andre, C., F. Boeri, and J. Marin (1976). "Synthese et realisation des systemes logiques a evolutions simultanees," *Rev. Fr. Autom. Inf. Rech. Oper.*, 10(4), pp. 67-86.

[Aslam 88] Aslam, S. (1988). *Use of Physical Simulation in the Instruction of Flexible Manufacturing Systems*, Master's Thesis, Rensselaer Polytechnic Institute, Troy, New York.

[Atabakche 86] Atabakche, H., D. S. Barbalho, R. Valette, and M. Courvoisier (1986). "From Petri net based PLCs to knowledge based control," in *Proc. of IECON*, pp. 812-817.

[Balbo 88] Balbo, G., S. C. Bruell, and S. Ghanta (1988). "Combining queuing networks and generalized stochastic Petri nets for the solution of complex models of system behavior," *IEEE Trans. on Computers*, 37(10), pp. 1251-1268.

[Balbo 87] Balbo, G., G. Chiola, G. Franceschinis, and G. M. Roet (1987). "Generalized stochastic Petri nets for the performance evaluation of FMS," in *Proc. IEEE Int. Conf. on Robotics and Automation*, Raleigh, North Carolina, pp. 1013-1018.

[Banaszak 90] Banaszak, Z. A. and B. H. Krogh (1990). "Deadlock avoidance in flexible manufacturing systems with concurrency competing process flows," *IEEE Trans. Robotics Automation*, 6(6), pp. 724-734.

[Beck 86] Beck, C. L. and B. H. Krogh (1986). "Models for simulation and discrete control of manufacturing systems," in *Proc. IEEE Int. Conf. on Robotics and Automation*, San Francisco, CA, pp. 305-310.

[Berge 62] Berge, C. (1962). *The Theory of Graphs,* John Wiley and Sons, NY.

[Berthelot 76] Berthelot, G. and G. Roucairol (1976). "Reduction of Petri nets," in *Mathematical Foundations of Computer Science*, A. Mazurkiewicz (ed.), Vol. 45, Springer-Verlag, pp. 202-209.

[Berthelot 86] Berthelot, G. (1986). "Checking properties of nets using transformations," in *Advances in Petri Nets 1985*, G. Rozenberg (ed.), Springer-Verlag, pp. 19-40.

[Berthelot 87] Berthelot, G. (1987). "Transformations and decompositions of nets," in *Advances in Petri Nets 1986*, W. Brauer, W. Reisig, and G. Rozenberg (eds.), Vol. 255, Part I, Springer-Verlag, pp. 359-376.

[Bollinger 88] Bollinger, J. G. and N. A. Duffie (1988). *Computer Control of Machines and Processes*, Addison-Wesley, pp. 369-409.

[Brand 88] Brand, K. and J. Kopainsky (1988). "Principles and engineering of process control with Petri nets," *IEEE Trans. on Automatic Control*, 33(2), pp. 138-149.

[Bruno 71] Bruno, J. and A. M. Altman (1971). "A theory of asynchronous control networks," *IEEE Trans. on Computer*, 20(6), pp.629-638.

[Bruno 85] Bruno, G. and P. Biglia (1985). "Performance evaluation and validation of tool handling in flexible manufacturing systems using Petri nets," *Proc. IEEE 1985 Int. Workshop on Timed Petri Nets*, Torino, Itaiy, pp. 64-71.

[Bruno 86] Bruno, G. and G. Marchetto (1986). "Process-translatable Petri nets for the rapid prototyping of process control systems," *IEEE Trans. on Software Engineering*, 12(2), pp. 346-357.

[Campos 90] Campos, J., J. M. Colom and M. Silva (1990). "Performance evaluation of repetitive automated manufacturing systems," in *Rensselaer's 2nd Int. Conf. on Computer Integrated Manufacturing*, Troy, NY, pp. 74-81.

[Chang 90] Chang, S. J. (1990). *Machine Intelligence and Learning Theories for Error Diagnosis and Recovery in Manufacturing Systems*, Doctoral Dissertation, ECSE, Rensselaer Polytechnic Institute, Troy, NY.

[Chang 91] Chang, S. J., F. DiCesare, and G. Goldbogen (1991). "Failure propagation trees for diagnosis in manufacturing systems," *IEEE Trans. on Systems, Man, and Cybernetics*, 21(4), pp. 767-776.

[Chao 92] Chao, D. Y., M. C. Zhou, and D. T. Wang (1992). "Extending knitting technique to Petri net synthesis of automated manufacturing systems," in *Proc. Rensselaer's Third Int. Conf. on Computer Integrated Manufacturing*, Troy, NY, pp. 56-63.

[Chen 91] Chen, Y. and C. Wongladkown (1991). "A real-time control simulator design for automated manufacturing systems using Petri nets," in *Proc. of IEEE Int. Conf. on Robotics and Automation,* Sacramento, CA, pp. 2542-2547.

[Chiola 85] Chiola, G. (1985). "A software package for the analysis of generalized stochastic Petri net models" in *Proc. IEEE Int. Workshop on Timed Petri Nets*, Torino, Italy, July, 1985.

[Chiola 87] Chiola, G. (1987). "A graphic Petri net tool for performance analysis," in *Proc. of Int. Workshop on Modeling Techniques and Performance Evaluation*, France, pp. 323-333.

[Ciardo 89] Ciardo, G. (1989). *Manual for the SPNP Package*, Duke University, February 1989.

[Ciardo 91] Ciardo, G. and K. S. Trivedi (1991). "A decomposition approach for stochastic Petri net models," in *Proc. of the 4th Int. Workshop on Petri Nets and Performance Models*, Melbourne, Australia, pp. 74-83.

[Cohen 85] Cohen, G., D. Dubois, J. P. Quadrat, and M. Viot (1985). "A linear-system-theoretic view of discrete-event processes and its use for performance evaluation in manufacturing, " *IEEE Trans. on Automatic Control*, 30(3), pp. 210-220.

[Colom 86] Colom, J. M., M. Silva, and J. L. Villarroel (1986). "On software implementation of Petri nets and colored Petri nets using high-level concurrent language," in *Proc. of the 7th European Workshop on Application and Theory of Petri Nets*, Oxford, pp. 207-241.

[Colom 87] Colom, J. M., J. Martinez and M. Silva (1987). "Packages for validating discrete production systems modeled with Petri nets," in *Applied Modeling and Simulation of Technological Systems*, P. Borne and S. G. Tzafestas (eds.), North-Holland, pp. 529-536.

[Commoner 71] Commoner, F., A. W. Holt, S. Even, and A. Pnueli (1971). "Marked directed graphs," *J. Comp. Syst. Sci.*, 5, pp. 511-532.

[Commoner 72] Commoner, F. (1972). "Deadlocks in Petri nets," *Report CA-7206-2311*, Massachusetts Computer Associates, Wakefield, MA.

[Coolahan 85] Coolahan, J. E. Jr. and N. Roussopoulos (1985). "A Timing requirements for time-driven systems using Augmented Petri nets," *IEEE Trans. on Software Engineering,* 9(5), pp. 603-616.

[Courvoisier 83] Courvoisier, M., R. Valette, J.M. Bigou, and P. Esteban (1983). "A programmable logic controller based on a high level specification tool," in *Proc. of IECON,* San Francisco, CA, pp. 174-179.

[Crockett 87] Crockett, D., A. Desrochers, F. DiCesare, and T. Ward (1987). "Implementation of a Petri net controller for a machining workstation," in *Proc. of IEEE Int. Conf. Robotics and Automation* Raleigh, NC, pp. 1861-1867.

[Cruette 92] Cruette, D., and J. C. Gentina (1992). "Design and validation of operation sequences in F.M.S," in *Proc. Rensselaer's Third Int. Conf. on Computer Integrated Manufacturing*, Troy, NY, pp. 262-268.

[Cumani 85] Cumani, A. (1985). "ESP-a package for evaluation of stochastic Petri nets with phase-type distributed transition times," in *Proc. IEEE Int. Workshop on Timed Petri Nets,* Torino, Italy, pp. 144-151.

[Datta 84a] Datta, A., D. Harms, and S. Ghosh (1984). "Deadlock avoidance in real-time resource sharing distributed systems: an approach using Petri nets," in *Proc. Int. Symp. on Real-time Systems,* pp. 49-61.

[Datta 84b] Datta, A. and S. Ghosh (1984). "Synthesis of a class of deadlock-free Petri nets," *Journal of the ACM,* 31(3), pp. 486-506.

[Datta 86] Datta, A. and S. Ghosh (1986). "Modular synthesis of deadlock-free control structures," *Foundations of Software Technology and Theoretical Computer Science*, Vol. 241, G. Goos and J. Hartmanis (ed,), Spinger-Verlag, pp. 288-318.

[Desrochers 89] Desrochers, A. A. (1989). *Modeling and Control of Automated Manufacturing Systems,* IEEE Computer Society Press.

[DiCesare 91] DiCesare, F. and A. A. Desrochers (1991). "Modeling, control, and performance analysis of automated manufacturing systems using Petri nets," in *Control and Dynamic Systems,* C. T. Leondes (ed.), Vol. 47, Academic Press, pp. 121-172.

[DiCesare 92] DiCesare, F. and M. C. Zhou (1992). "Symbolic performance evaluation of concurrent systems combining Petri nets and moment generating functions," To appear in *Control and Dynamic Systems,* C. T. Leondes (ed.), Academic Press.

[Dijkstra 68] Dijkstra, E. W. (1968). "Co-operating sequential processes," in *Programming Languages,* F. Genuys (ed.), New York: Academic, pp. 43-112.

[Dubois 83] Dubois, D. and K. Stecke (1983). "Using Petri nets to represent production processes," in *Proc. of the 22nd IEEE Conf. on Decision and Control,* San Antonio, TX, pp. 1062-1067,

[Dugan 85] Dugan, J. B., A. Bobbio, A. Ciardo, and K. S. Trivedi (1985). "The design of a unified package for the solution of stochastic Petri net models," in *Proc. 1985 Int. Workshop on Timed Petri nets,* Torino, Italy, pp. 6-13.

[Dunkler 88] Dunkler, O., C. M. Mitchell, T. Govindaraj, and J. C. Ammons (1988). "The effectiveness of supervisory control strategies in scheduling flexible manufacturing systems," *IEEE Trans. on Systems, Man, and Cybernetics,* 18(2), pp. 223-237.

[Eleftheriu 89] Eleftheriu, M. N. (1989). *On the Analysis and Control of Multistage Production Manufacturing Systems,* Doctoral Dissertation, Rensselaer Polytechnic Institute, Troy, NY.

[Ezpeleta 92] Ezpeleta, J., and J. Martinez (1992). "Formal Specification and Validation in Production Plants" in *Proc. Rensselaer's Third Int. Conf. on Computer Integrated Manufacturing,* Troy, NY, pp. 64-73.

[Feicht 87] Feicht, D., F. DiCesare and G. Goldbogen (1987). "Intelligent error recovery for manufacturing workstations," in *Proceedings of the Tenth IASTED International Symposium on Robotics and Automation,* Lugano, Switzerland, pp. 119-123.

[Ferrarini 92] Ferrarini, L. (1992). "An incremental approach to logic controller design with Petri nets," *IEEE Trans. on System, Man, and Cybernetics* 22(3), pp. 461-473.

[Fielding 88] Fielding, P. J., F. DiCesare, and G. Goldbogen (1988) "Error recovery in automated manufacturing through the augmentation of programmed processes," *Journal of Robotic Systems,* 5(4), pp. 337-362.

[Fielding 89] Fielding, P. J. (1989). *Learning Error Recoveries in Automated Manufacturing by General Explanation of Observed Program Augmentations,* Doctoral Dissertation, ECSE, Rensselaer Polytechnic Institute, Troy, NY.

[Florin 82] Florin, G. and S. Natkin, "Evaluation based upon stochastic Petri nets of the maximum throughput of a full dulex protocol," in *Informatik Fachberichte,* C. Girault and W. Reisig (eds.), Springer-Verlag, 1982.

[Freedman 88] Freedman, P. and A. Malowany (1988). "Petri nets and modeling of repetitive events," *TR-CIM-88-5,* McGill Research Center for Intelligent Machines, McGill University, Quebec, Canada.

[Freedman 91] Freedman, P. (1991). "Time, Petri nets, and robotics," *IEEE Trans. on Robotics and Automation,* 7(4), pp. 417-433.

[Genrich 80] Genrich, H., K. Lautenbach, and P. S. Thiagarajan (1980). "Elements of general net theory," *Net Theory and Applications,* Vol. 84, W. Brauer (ed.), Springer Verlag, pp. 121-163.

[Genrich 81] Genrich, H. and K. Lautenbach (1981). "System modeling with high level Petri nets," *Theoretical Computer Science*, 13, pp. 109-136.

[Gentina 87] Gentina, J. C. and D. Corbeel (1987). "Colored adaptive structured Petri net: a tool for the automated synthesis of hierarchical control of flexible manufacturing systems," in *Proc. of IEEE Int. Conf. on Robotics and Automation,* Raleigh, NC, pp. 1166-1173.

[Gershwin 86] Gershwin, S. G., R. R. Hildebrant, R. Suri, and S. K. Mitter (1986). "A control perspective on recent trends in manufacturing systems," *IEEE Control Systems Magazine,* pp. 3-15.

[Gini 83] Gini, M. and G. Gini (1983). "Towards automatic error recovery in robot program," in *Proc. of the Eighth Int. Joint Conf. on Artificial Intelligence,* Karlsruhe, Germany, pp 821-823.

[Giua, 91] Giua, A. (1991). *Petri Nets as Discrete Event Models for Supervisory Control*, Doctoral Dissertation, ECSE, Rensselaer Polytechnic Institute, Troy, NY.

[Goltz 82] Goltz, U., W. Reisig, and P. S. Thiagarajan (1982). "Two alternative definitions of synchronic distance," in *Application and Theory of Petri Nets*, C. Girault and W. Resisg (eds.), Springer Verlag, pp. 96-103.

[Goltz 86] Goltz, U. and C. Y. Yuan (1986). "Synchronic structure," in *Advances in Petri Nets,* Vol. 222, W. Brauer (ed.), Springer Verlag, pp. 233-252.

[Gressier 85] Gressier, E. (1985). "A stochastic Petri net model for Ethernet," in *Proc. 1985 Int. Workshop on Timed Petri nets,* Torino, Italy, pp. 296-303.

[Guo 91] Guo, D. L., F. DiCesare, and M. C. Zhou (1991). "Moment generating function approach to performance analysis of extended stochastic Petri nets," in *Proc. of IEEE Int. Conf. on Robotics and Automation,* Sacramento, CA, pp. 1309-1314.

[Guo 92] Guo, D. L., F. DiCesare, and M. C. Zhou (1992). "A moment generating function based approach for evaluating extended stochastic Petri nets," *IEEE Trans. on Automatic Control,* 37(11).

[Hack 72] Hack, M. (1972). *Analysis of Production Schemata by Petri Nets.* MS Thesis, EE Dept., MIT Cambridge, MA.

[Hack 75] Hack, M. (1975). *Decidability Question for Petri Nets.* Doctoral Dissertation, EE Dept., MIT, Cambridge, MA.

[Hauschildt 85] Hauschildt, D. and R. Valk (1985). "Safe states in banker like resource allocation problems," in *Advances in Petri Nets,* Vol. 222, W. Brauer (ed.), Springer Verlag, pp. 253-277.

[Hillion 89] Hillion, H. P. and J. M. Proth, "Performance evaluation of job-shop systems using timed-event-graphs," *IEEE Trans. on Automatic Control,* 34(1), pp. 3-9, 1989.

[Ho 79] Ho, Y. C., M. A. Eyler, and T. T. Chien (1979) "A gradient technique for general buffer storage design in a production line," *Int. J. of Production Research,* 17(6), pp. 557-580.

[Ho 87] Ho, Y. C. (1987). "Performance evaluation and perturbation analysis of discrete event dynamic systems," *IEEE Trans. on Automatic Control,* 32(7), pp. 563-572.

[Hoare 85] Hoare, C. A. R. (1985). *Communicating Sequential Processes,* Prentice-Hall, Englewood Cliffs, N.J.

[Holliday 85] Holliday, M. A. and M. K. Vernon (1985). "A generalized timed Petri net model for performance analysis," in *Proc. 1985 IEEE Int. Workshop on Timed Petri Nets,* Torino, Italy, pp. 181-190.

[Holt 70] Holt, A. W. and F. Commoner (1970). *Events and Conditions,* Applied Data Research, NY.

[Hsieh 92] Hsieh, D.-Y. and S.-C. Chang (1992). "Deadlock avoidance controller synthesis for flexible manufacturing systems," in *Proc. Rensselaer's Third Int. Conf. on Computer Integrated Manufacturing,* Troy, NY, pp. 252-261.

[Inan 88] Inan, K. and P. Varaiya (1988). "Finitely recursive process models for discrete event systems," *IEEE Trans. on Automatic Control,* 33(7), pp. 626-639.

[Jacob 88] Jacob, R. T. and I. P. Page (1988). "Synthesis of mutual exclusion solutions based on binary semaphores," *IEEE Trans. on Software Engineering,* 15(5), pp. 560-568.

[Jafari 92] Jafari, M. A. (1992). "An architecture for a shop-floor controller using colored Petri nets," *Int. J. of Flexible Manufacturing Systems,* 4, pp. 159-181.

[Jeng 90] Jeng, M. D. and F. DiCesare (1990). "A review of synthesis techniques for Petri nets," in *Proc. of Rensselaer's 2nd Int. Conf. on Computer Integrated Manufacturing,* Troy, NY, pp. 348-355.

[Jeng 92a] Jeng, M. D. (1992). *Theory and Applications of Resource Control Petri Nets for Automated Manufacturing Systems,* Doctoral Dissertation, ECSE, Rensselaer Polytechnic Institute, Troy, NY.

[Jeng 92b] Jeng, M. D. and Frank DiCesare (1992). "A modular Petri net synthesis approach for automated manufacturing systems," in *Proc. Japan U.S.A. Symp. on Flexible Automation*, San Francisco, CA, pp. 1163-1170.

[Jensen 81a] Jensen, K. (1981). "Colored Petri nets and the invariant method," *Theoretical Computer Science*, 14, pp. 317-336.

[Jensen 81b] Jensen, K. (1981). "How to find invariants for colored Petri nets," in *LNCS: Math. Foundation Comput. Sci.*, Vol. 118, pp. 327-338.

[Jensen 86] Jensen, K. (1986). "Colored Petri nets," in *Lecture Notes in Computer Science Vol. 254: Advances in Petri Nets 1986,* part 1, Springer-Verlag, pp. 248-299.

[Jonsonbaugh 81] Jonsonbaugh, R. and T. Murata (1981). "Additional methods for reduction and expansion of marked graphs," *IEEE Trans. on Circuit Systems,* 28(10), pp. 1009-1014.

[Jungnitz 92] Jungnitz, H. J. (1992). *Approximation Methods for Stochastic Petri Nets,* Doctoral Dissertation, ECSE, Rensselaer Polytechnic Institute, Troy, NY.

[Jungnitz 91] Jungnitz, H. J. and A. A. Desrochers (1991). "Flow equivalent nets for the performance analysis of generalized stochastic Petri nets," in *Proceedings of the IEEE Int. Conf. on Robotics and Automation,* Sacramento, CA, pp. 122-127.

[Karp 69] Karp, R. M. and R. E. Miller (1969). "Parallel program schemata," *J. of Comp. & Sys. Sci.*, 3, pp. 147-195.

[Kasturia 88] Kasturia, E., F. DiCesare, and A. A. Desrochers (1988). "Real-time control of multilevel manufacturing systems using colored Petri nets," in *Proc. of IEEE Int. Conf. Robotics and Automation*, Phildephia, PA, pp. 1114-1119.

[Koh 90] Koh, I. and F. DiCesare (1990). "Transformation methods for generalized Petri nets and their applications in flexible manufacturing systems," in *Rensselaer's 2nd Int. Conf. on Computer Integrated Manufacturing,* Troy, NY, pp. 364-371.

[Koh 91a] Koh, I. and F. DiCesare (1991). "Modular transformation methods for generalized Petri nets and their applications in manufacturing automation," *IEEE Trans. on Systems, Man, and Cybernetics,* 21(6), pp. 963-973.

[Koh 91b] Koh, I. (1991). *A Transformation Theory for Petri Nets and their Applications to Manufacturing Automation*, Doctoral Dissertation, ECSE, Rensselaer Polytechnic Institute, Troy, NY.

[Krogh 86] Krogh, B. H. and C. L. Beck (1986). "Synthesis of place/transition nets for simulation and control of manufacturing Systems," in *Preprints of 4th IFAC/IFORS Symp. Large Scale Systems,* Zurich, pp. 661-666.

[Krogh 87] Krogh, B. H. and R. S. Sreenivas (1987). "Essentially decision free Petri nets for real-time resource allocation," in *Proc. of IEEE Int. Conf. Robotics and Automation,* Raleigh, NC, pp. 1005-1011.

[Krogh 88] Krogh, B. H., R. Willson, and D. Pathak (1988). "Automated generation and evaluation of control programs for discrete manufacturing processes," in *Rensselaer's 1st. Int. Conf. on Computer Integrated Manufacturing,* Troy, NY, pp. 92-99.

[Kumaradjaja 89] Kumaradjaja, R. (1989). *A Causal Reasoning Model for Plan Generation, Execution Monitoring and Error Recovery in Automated Manufacturing Systems*, Doctoral Dissertation, ECSE, Rensselaer Polytechnic Institute, Troy, NY.

[Lautenbach 75] Lautenbach, K. (1975). "Liveness in Petri nets," Internal Report, Gesellschaft fur Mathematik und Datenverarbeitung, GMD-ISF 02.1/75-7-29.

[Lautenbach 79] Lautenbach, K. and P. S. Thiagarajan (1979). "Analysis of a resource allocation problem using Petri nets," in *First European Conf. on Distributed Processing*, Toulouse, pp. 260-266.

[Lautenbach 87] Lautenbach, K. (1987). "Linear algebraic techniques for place/transition nets," in *Advances in Petri Nets 1986,* W. Brauer, W. Reisig, and G. Rozenberg (eds.), Springer-Verlag, NY, pp. 142-167.

[Lee 83] Lee, M., D. Barnes and N. Hardy (1983). "Knowledge base error recovery in industrial robots," in *Proc. of the Eighth Int. Joint Conf. on Artificial Intelligence*, pp. 824-826.

[Lee 85] Lee, K. H. and J. Favrel (1985). "Hierarchical reduction method for analysis and decomposition of Petri nets," *IEEE Trans. on Systems, Man, and Cybernetics,* 15(2), pp. 272-280.

[Lee 87] Lee, K. H., J. Favrel, and P. Baptiste (1987). "Generalized Petri net reduction method," *IEEE Trans. on Systems, Man, and Cybernetics,* 17(2), pp. 297-303.

[Lee 92] Lee, D. Y. and Frank DiCesare (1992). "Experimental study of a heuristic function for FMS scheduling," in *Proc. Japan-U.S.A. Symposium on Flexible Automation*, San Francisco, CA, pp. 1171-1178.

[Ma 92] Ma, J. and M. C. Zhou (1992). "Performance evaluation of discrete event systems via stepwise reduction and approximation of stochastic Petri nets," To appear in *Proc. 31st IEEE Int. Conf. on Decision and Control,* Tucson, AZ.

[Magott 85] Magott, J. (1985). "Performance evaluation of systems of cyclic sequential processes with mutual exclusion using Petri nets," *Information Processing Letters,* 21, 229-232.

[Magott 87] Magott, J. (1987). "Performance evaluation of systems of cyclic sequential processes with mutual exclusion and communication by buffers using timed Petri nets," in *Proc. of 1987 Int. Workshop on Petri Nets and Performance Models,* Madsion, WI, pp. 146-153.

[Martinez 82] Martinez, J. and M. Silva (1982). "A simple and fast algorithm to obtain all invariants of a generalized Petri nets," in *Application and Theory of Petri Nets,* Informatic Fachbericht 52, Springer-Verlag, pp. 301-310.

[Martinez 86] Martinez J., H. Alla, and M. Silva (1986). "Petri nets for the specifications of FMSs," in *Modeling and Design of Flexible Manufacturing Systems*, A. Kusiak (ed.), Elsevier Science Publishers, Amsterdam, pp. 389-406.

[Martinez 87] Martinez, J., P. Muro, and M. Silva (1987). "Modeling, validation and software implementation of production systems using high level Petri nets," in *Proc. of IEEE Int. Conf. on Robotics and Automation,* Raleigh, North Carolina, pp. 1180-1185.

[Martinez 88] Martinez, J., P. R. Muro, M. Silva, S.F. Smith and J. L. Villarroel (1988). "Merging artificial intelligence techniques and Petri nets for real time scheduling and control of production systems," in *Proc. of 12th IMACS World Congress on Scientific Computation*, Paris, Vol. 3, pp. 528-531.

[Menon 88] Menon, S. R., T. J. Quinn, P. M. Ferreira and S. G. Kapoor (1988). "Coordination control of flexible manufacturing systems using colored Petri nets," in *Proc. of Int. Conf. on Computer Aided Production Engineering*, pp. 317-326.

[Merabet 86] Merabet, A. (1986). "Synchronization of operations in a flexible manufacturing cell: the Petri net approach," *J. of Manufacturing Systems*, 5(3), pp. 161-169.

[Merlin 76a] Merlin, P. and D. Farber (1976). "On the recovery of communication protocols," in *Proc. of 1976 Int. Conf. on Communications,* New York, pp. 20-26.

[Merlin 76b] Merlin, P. and D. Farber (1976). "A methodology for the design and implementation of communication protocols," *IEEE Trans. on Communications,* 24(6), pp. 614-621.

[Molloy 82] Molloy, M. K. (1982). "Performance analysis using stochastic Petri nets," *IEEE Trans. on Computers,* 31(9), pp. 913-917.

[Morganti 82] Morganti, M. P. (1982). "Petri net implementation of recovery strategies in a large ESS," in *Application and Theory of Petri Nets*, C. Girault and W. Reisig (eds.), Informatik-Fachberichte, 52, Springer-Verlag, NY, pp. 212-221.

[Murata 77] Murata, T. (1977). "Circuit theoretic analysis and synthesis of marked graphs," *IEEE Trans. Circ. and Sys.*, 27, pp. 400-405.

[Murata 80a] Murata, T. (1980). "Synthesis of decision-free concurrent systems for prescribed resources and performance," *IEEE Trans. Software Engineering,* 6(6), pp. 525-530.

[Murata 80b] Murata, T. and J. Y. Koh (1980). "Reduction and expansion of live and safe marked graphs," *IEEE Trans. Circ. and Sys.*, 27, pp. 68-70.

[Murata 86] Murata, T., N. Komoda, and K. Matsumoto (1986). "A Petri net based controller for flexible and maintainable sequence control and its applications in factory automation," *IEEE Trans. on Industrial Electronics*, 33, pp. 1-8.

[Murata 89] Murata, T. (1989). "Petri nets: properties, analysis and application," *Proc. of the IEEE,* 77(4), pp. 541-579.

[Narahari 85] Narahari, Y. and N. Viswanadham (1985). "A Petri net approach to the modeling and analysis of flexible manufacturing systems," *Annals of Operations Research,* 3, pp. 449-472.

[Narahari 86] Narahari, Y. and N. Viswanadham (1985). "On the Invariants of Colored Petri Nets," in *Advances in Petri Nets 1985*, Vol. 222, pp. 330-345.

[Narahari 87] Narahari, Y. and N. Viswanadham (1987). "Colored Petri nets: a tool for modeling, validation, and simulation of FMS," in *Proc. of IEEE Int. Conf. on Robotics and Automation,* Raleigh, NC, pp. 1985-1990.

[Naylor 87] Naylor, A. W. and R. A. Volz (1987). "Design of integrated manufacturing system control software," *IEEE Trans. on Systems, Man, and Cybernetics,* 17(6), pp. 881-897.

[Patil 70] Patil, S. S. (1970), *Coordination of Asynchronous Events.* Doctoral Dissertation, MIT, Cambridge, MA.

[Peterson 81] Peterson, J. L. (1981). *Petri Net Theory and the Modeling of Systems*, Prentice, Inc., Englewood Cliffs, NJ.

[Petri 62] Petri, C. A. (1962). *Communication with Automata.* Doctoral Dissertation, University of Bonn, Bonn, West Germany.

[Petri 79] Petri, C. A. (1979). "Introduction to general net theory," in *Lecture Notes in Computer Science: Net Theory and Application,* pp. 1-19.

[Ramadge 83] Ramadge, P. J. and W. M. Wonham (1983). "Supervisory control of a class of discrete event processes," *System Control Group Report 8311*, Dept. of Electrical Engineering, Uni. of Toronto, Dec. 1983; *SIAM J. Control Optim.* 25(1), pp. 206-230, 1987.

[Ramadge 89] Ramadge, P. J. G. and W. M. Wonham, "The control of discrete event systems," *Proc. IEEE*, 77(1), pp. 81–98.

[Ramamoorthy 80] Ramamoorthy, C. V. and G. S. Ho (1980). "Performance evaluation of asynchronous concurrent systems using Petri nets," *IEEE Trans. on Software Engineering,* 6(5), pp. 440-449.

[Ramchandani 74] Ramchandani, C. (1974). *Analysis of Asynchronous Concurrent Systems by Timed Petri Nets,* Doctoral Dissertation, MIT, Cambridge, MA.

[Reisig 85] Reisig, W. (1985). *Petri Nets: An Introduction,* Spring-Verlag.

[Rozenberg 86] Rozenberg, G. and P. S. Thiagarajan (1986). "Petri nets: basic notions, structure, and behavior," *Current Trends in Concurrency,* J. W. de Bakker et al. (ed.), Springer-Verlag, pp. 585-668.

[Rudolph 89] Rudolph, D. L. (1989). *Petri Net-based Control of a Flexible Manufacturing System,* Master's Thesis, Rensselaer Polytechnic Institute, Troy, NY.

[Sahraoui 86] Sahraoui, A., M. Courvoisier and R. Valette (1986). "Some considerations on monitoring in distributed real-time control of flexible manufacturing systems," in *Proc. of IECON 86,* pp. 805-810.

[Sahraoui 87] Sahraoui, A., H. Atabakhche, M. Courvoisier, R. Valette (1987). "Joining Petri nets and knowledge based systems for monitoring purposes," in *Proc. of IEEE Int. Conf. on Robotics and Automation,* Raleigh, NC, pp. 1861-1867.

[Seidmann 89] Seidmann, A. and S. Y. Nof (1989). "Operational analysis of an autonomous assembly robotic station," *IEEE Trans. on Robotics and Automation,* 5(1), pp. 4-15.

[Shukla 91d] Shukla, A. and A. D. Robbi (1991). "A Petri net simulation tool," in *Proc. 1991 IEEE Int. Conf. on Systems, Man, and Cybernetics,* Charlottesville, VA, pp. 361-366.

[Sifakis 77] Sifakis, J. (1977). "Use of Petri nets for performance evaluation," in *Measuring, Modelling, and Evaluating Computer Systems,* Amsterdam: North-Holland, pp. 75-93.

[Sifakis 78] Sifakis, J. (1978). "Structural properties of Petri nets," in *Mathematical Foundation of Computer Science,* Vol. 64, J. Winkowski (ed.), Springer-Verlag, pp. 474-483.

[Sifakis 79] Sifakis, J. (1979). "Realization of fault-tolerant systems by coding Petri nets," *J. of Design Automation and Fault Tolerant Computing,* 3(1), pp. 93-107.

[Silva 81] Silva, M. (1981). "Sur le concept de macro place et son utilisation pour l'analyse des reseaux de Petri," *PAIRO Automat.*, 15(4), pp. 335-345.

[Silva 82] Silva, M. and S. Velilla (1982). "Programmable logic controller and Petri nets: a comparative study," in *IFAC Conf. on Software for Computer Control,* Madrid, Spain. pp. 83-88.

[Silva 85a] Silva, M. (1985). *Las redes de Petri en la Automatica y la Informatica,* Editorial AC, Madrid.

[Silva 85b] Silva, M. and S. Velilla (1985). "Error detection and correction on Petri net models of discrete event control systems," in *Proceeding of ISCAS 85,* pp. 921-924.

[Silva 82] Silva, M. (1987). "Toward a synchrony theory for P/T nets," *Concurrency and Nets,* K. Voss, *et al.* (ed.), Springer-Verlag, pp. 435-460.

[Silva 90] Silva, M. and R. Valette (1990). "Petri nets and flexible manufacturing," in *Advances in Petri Nets 1989,* G. Rozenberg (ed.), Springer-Verlag, pp. 374-417.

[Singhal 89] Singhal, M. (1989). "Deadlock detection in distributed systems," *IEEE Computer Magazine,* pp. 37-48.

[Srinivas 78] Srinivas, S. (1987). "Error recovery in robots through failure reasoning analysis," in *AFIC Proceedings of National Computer Conference,* Anaheim, CA, pp. 275-282.

[Suraj 80] Suraj, Z. (1980). "A resource allocation problem," in *Mathematical Foundation of Computer Science,* P. Dembinski (ed.), Vol. 88, Springer Verlag, pp. 627-642.

[Suzuki 80] Suzuki, I. and T. Murata (1980). "A method for hierarchically representing large scale Petri nets," in *Proc. of the 1980 Int. Conf. on Circuits and Computers,* pp. 620-623.

[Suzuki 83a] Suzuki, I. and T. Murata (1983a). "A method for stepwise refinements and abstractions of Petri nets," *J. of Comp. & Syst. Sci.,* 27, pp. 51-76.

[Suzuki 83b] Suzuki, I. and T. Kasami (1983b). "Three measures for synchronic dependence in Petri nets," *Acta Informatica,* 19, pp. 325-338.

[Tang 92] Tang, T. N. and M. C. Zhou (1992). "Development and applications of a Petri net simulation tool for manufacturing systems," in *Proc. Japan U.S.A. Symp. on Flexible Automation,* San Francisco, CA, pp. 1199-1205.

[Tazza 87] Tazza, M. (1987). "Quantitative analysis of a resource allocation problem: a net theory based proposal," in *Concurrency and Nets,* K. Voss, H. J. Genrich, and G. Rozenberg (eds.), Springer-Verlag, pp. 511-532.

[Valavanis 90] Valvanis, K. S. (1990). "On the hierarchical analysis and simulation of flexible manufacturing systems with extended Petri nets," *IEEE Trans. on System, Man, and Cybernetics,* 20 (1), pp. 94-100.

[Valette 79] Valette, R. (1979). "Analysis of Petri nets by stepwise refinements," *J. of Comp. and Syst. Sci.,* 18, 35-46.

[Valette 82a] Valette, R., M. Courvoisier, and D. Mayeux (1982). "Control of flexible production systems and Petri nets," in *Informatik Fachberichte 66,* Springer Verlag, pp. 264-267.

[Valette 82b] Valette, R. (1982). "Petri nets and reliable real-time systems," in *Application and Theory of Petri Nets,* C. Girault and W. Reisig (eds.), Informatik-Fachberichte, 52, Springer-Verlag, pp. 212-221.

[Valette 85a] Valette, R., M. Courvoisier, H. Demmou, J. M. Bigou, and C. Desclaux (1985). "Putting Petri nets to work for controlling flexible manufacturing systems," in *Proc. of Int. Symp. on Circ. & Sys.,* Kyoto, Japan, pp. 929-932.

[Valette 85b] Valette, R.,V. Thomas, and S. Bachmann (1985). "SEDRIC: un simulateur a evenements discrets base sur les reseaux de Petri," *Systemes de production,* APII, 19(5), pp. 423-436.

[Valette 87] Valette, R. (1987). "Nets in production systems," in *Advances in Petri Nets 1986,* W. Brauer, W. Reisig, and G. Rozenberg (eds.), Vol. 255, Part I, Springer-Verlag, pp. 191-217.

[Velilla 88] Velilla, S. and M. Silva (1988). "The SPY: a mechanism for safe implementation of highly concurrent systems," in *15th IFAC/IFIP Workshop on Real-Time Programming,* Valencia, Spain, pp. 95-101.

[Villarroel 88] Villarroel, J. L., J. Martinez, M. Silva (1988). "GRAMAN: a graphic system for manufacturing system design," in *IMACS Int. Symp. on Sys. Model. & Simul.* (SMS'88), Cetraro, Italy, pp. 311-316.

[Viswanadham 87] Viswanadham, N. and Y. Narahari (1987). "Colored Petri net models for automated manufacturing systems," in *Proc. of IEEE Int. Conf. on Robotics and Automation,* Raleigh, North Carolina, pp. 1985-1990,

[Viswanadham 88a] Viswanadham, N. and Y. Narahari (1988). "Stochastic Petri net models for performance evaluation of automated manufacturing systems" *Information and Decision Technologies,* 14, North-Holland, pp. 125-142.

[Viswanadham 88b] Viswanadham, N. and T. L. Johnson (1988). "Fault detection and diagnosis of automated manufacturing systems," in *Proc. of 27th on IEEE Int. Conf. Decision and Control*, pp. 2031-2306.

[Viswanadham 90] Viswanadham, N. and T. L. Johnson (1990). "Performance analysis of automated manufacturing systems with blocking and deadlock," in *Rensselaer's 2nd Int. Conf. on Computer Integrated Manufacturing,* Troy, NY, pp. 64-68.

[Wang 90a] Wang, F. Y. (1990). *A Coordination Theory of Intelligent Machines,* Doctoral Dissertation, ECSE, Rensselaer Polytechnic Institute, Troy, NY.

[Wang 90b] Wang, F. Y. and G. N. Saridis (1990). "A coordination theory for intelligent machines," *Automatica,* 26(5), pp. 833-844.

[Wang 89] Wang, F. Y. and K. Gildea (1989). "A colored Petri net model for connection management services in manufacturing message specification (MMS)," *Computer Communication Review,* 19(3), pp. 76-98.

[Waston 91] Waston III, J. F. and A. A. Desrochers (1991). "Applying GSPNs to manufacturing systems containing non-exponential transition functions," in *Proc. of IEEE Int. Conf. on Robotics and Automation,* Sacramento, CA, pp. 366-371.

[Weiss 86] Weiss, J. (1986). *Development of a Physical Simulator for a Flexible Manufacturing System,* Master's Thesis, Rensselaer Polytechnic Institute, Troy, NY.

[Wiendahl 85] Wiendahl, H. P. and W. Ziersch (1985). "Increasing the availability of assembly systems," *Assembly Automation,* 5(4), pp. 217-224.

[Wonham 84] Wonham, W. M. and P. J. Ramadge (1984). "On modular synthesis of supervisory controls for discrete-event processes," in *Proc. of the IEEE and I. I. Sc. Int. Conf. on Computers, Systems and Signal Processing,* Bangalore, pp. 500- 504.

[Wonham 87] Wonham, W. M. (1987). "A control theory for discrete-event systems," *System Control Group Report 8714,* Dept. of Electrical Engineering, Uni. of Toronto, Toronto, Canada.

[Wu 84] Wu, Z. H. and T. Murata (1984). "Use of Petri nets for distributed control of fairness in concurrent system," in *Proc. First Conf. Computers Applications,* Beijing, pp. 84-91.

[Yao 85] Yao, D. D. and J. A. Buzacott (1985). "Modeling the performance of flexible manufacturing systems," *J. Prod. Res.*, 23(5), pp. 945-959.

[Yoeli 80] Yoeli, M. (1980). "Synthesis of concurrent systems," in *Application and Theory of Petri Nets,* C. Girault and W. Reisig (ed.), Springer-Verlag, pp. 183-186.

[Yuan 86] Yuan, C. Y. (1986). "Synchronic distances in C/E-systems," in *Advances in Petri Nets*, Vol. 222, W. Brauer (ed.), Springer Verlag, pp. 101-121.

[Yuan 88] Yuan, C. Y. (1988). *Petri Nets*, Southeastern University Press, Nanjing, China.

[Zeng 91] Zeng, L. and H.-P. N. Wang (1991). "Conflict detection of automated guided vehicles: a Petri net approach," *Int. J. Production Research,* 29(5), pp. 856-879.

[Zhang 92] Zhang, D. (1992). "Planning using timed Pr/T Nets," in *Proc. Japan U.S.A. Symposium on Flexible Automation*, San Francisco, CA, pp. 1179-1184.

[Zhou 88] Zhou, M. C. and F. DiCesare (1988). "Adaptive design of Petri net controllers for automatic error recovery," in *Proc. IEEE Int. Symp. on Intelligent Control*, Arlington, VA, pp. 652-657.

[Zhou 89a] Zhou, M. C. and F. DiCesare (1989). "Adaptive design of Petri net controllers for error recovery in automated manufacturing systems," *IEEE Trans. on Systems, Man, and Cybernetics,* 19(5), pp. 963-973.

[Zhou 89b] Zhou, M. C., F. DiCesare, and A. A. Desrochers (1989). "A top-down modular approach to synthesis of Petri net models for manufacturing systems," in *Proc. of IEEE Int. Conf. Robotics and Automation,* Scottsdale, AZ, pp. 534-539.

[Zhou 89c] Zhou, M. C., F. DiCesare, and G. Goldbogen (1989). "Computer-aided design of discrete event dynamic systems," in *Proc. of 1989 Int. Conf. on CAD and Computer Graphics,* Beijing, China, pp. 766-771.

[Zhou 90a] Zhou, M. C. (1990). *A Theory for the Synthesis and Augmentation of Petri Nets in Automation,* Doctoral Dissertation, ECSE, Rensselaer Polytechnic Institute, Troy, NY.

[Zhou 90b] Zhou, M. C. and F. DiCesare (1990). "A Petri net design method for automated manufacturing systems with shared resources," in *Proc. of IEEE Int. Conf. on Robotics and Automation,* Cincinnati, OH, pp. 526-531.

[Zhou 90c] Zhou, M. C., F. DiCesare, and D. Rudolph (1990). "Control of a flexible manufacturing system using Petri nets," in *1990 Reprints of 11th IFAC Congress Conference,* Tallin, USSR, Vol. 9, pp. 43-48.

[Zhou 90d] Zhou, M. C., F. DiCesare, and D. Guo (1990). "Modeling and performance analysis of a resource-sharing manufacturing system using stochastic Petri nets," in *Proc. IEEE Int. Symp. on Intelligent Control.* Philadelphia, PA. pp. 1005-1010.

[Zhou 90e] Zhou, M. C. and F. DiCesare (1990). "Modeling buffers in automated manufacturing systems using Petri nets," in *Proc. of Rensselaer's Second Int. Conf. on Computer Integrated Manufacturing,* Troy, NY, pp. 265-272.

[Zhou 90f] Zhou, M. C., and F. DiCesare (1990). "Hybrid synthesis of Petri nets for manufacturing systems: a case study," in *Proc. of 3rd Int. Symp. on Robotics & Manufacturing (ISRAM90),* Vancouver, Canada, pp. 667-672.

[Zhou 91a] Zhou, M. C. and F. DiCesare (1991). "Parallel and sequential mutual exclusions for Petri net modeling for manufacturing systems," *IEEE Trans. on Robotics and Automation,* 7(4), pp. 515-527.

[Zhou 91b] Zhou, M. C. and M. C. Leu (1991). "Modeling and performance analysis of a flexible PCB assembly station using Petri nets," *Trans. of the ASME, Journal of Electronic Packaging,* 113(4), pp. 410-416.

[Zhou 91c] Zhou, M. C., F. Wang, and D. L. Guo (1991). "Deadlock avoidance in a five-robot-five-assembly-line system: Petri net modeling and analysis," in *Reprints of IFAC Workshop on Discrete Event System Theory and Applications in Manufacturing and Social Phenomena (91'DES),* Shenyang, China, pp. 123-128.

[Zhou 91d] Zhou, M. C., K. McDermott, P. A. Patel, and T. N. Tang (1991). "Construction of Petri net-based mathematical models of an FMS cell," in *Proc. 1991 IEEE Int. Conf. on Systems, Man, and Cybernetics,* Charlottesville, VA, pp. 367-372.

[Zhou 91e] Zhou, M. C. (1991). "Combination of Petri nets and intelligent decision makers for manufacturing systems control," in *Proc. of 6th IEEE Int. Symp. on Intelligent Control,* Arlington, VA, pp. 146-151.

[Zhou 92a] Zhou, M. C., F. DiCesare, and A. A. Desrochers (1992). "A hybrid methodology for synthesis of Petri nets for manufacturing systems," *IEEE Trans. on Robotics and Automation,* 8(3), pp. 350-361.

[Zhou 92b] Zhou, M. C., F. DiCesare, and D. Rudolph (1992). "Design and implementation of a Petri net supervisor for a flexible manufacturing system," to appear in *Automatica*, Vol. 28.

[Zhou 92c] Zhou, M. C. (1992). "Computer-aided modeling, analysis, and design of discrete Event Systems using Petri nets," in *Proc. of 1992 IEEE Symp. on Computer-Aided Control System Design (CACSD'92)*, Napa, CA, pp. 255-261.

[Zhou 92d] Zhou, M. C., D. L. Guo, and F. DiCesare (1992). "Integration of Petri nets and moment generating function approaches for system performance evaluation," To appear in *J. of Systems Integration.*

[Zhou 93] Zhou, M. C., K. McDermott, and P. A. Patel (1993). "Petri net synthesis and analysis of an FMS cell," to appear in *IEEE Trans. on Systems, Man, and Cybernetics*, 23(2).

[Zurawski 92] Zurawski, R. (1992). "A class of Petri net models used to represent primary components of flexible manufacturing systems," in *Proc. Japan U.S.A. Symposium on Flexible Automation*, San Francisco, CA, pp. 1695-1702.

Index